Solar Power Forecasting

This book takes an approach that leverages methods using time series analysis, machine learning, and stochastic models to effectively forecast solar power. The goal of this book is not only to produce an accurate forecast but also to make it conducive to be used for decision-making.

Solar Power Forecasting: Using Time Series and Machine Learning combines traditional forecasting with recent advances in machine learning and data science. It uses a decision-making-oriented approach and provides probabilistic forecasts and methods, as well as explains the analytical underpinnings of accuracy metrics in detail. As it illustrates through examples of how forecasting can be used in planning and operations, the book also delivers a systems-level approach.

The book is a single source of information for various aspects of solar forecasting such as data science methods, computational aspects, and mathematical foundations. It is useful for practitioners, students, and seasoned researchers not only in the solar power field but also for forecasting in general.

Color figures can be found on Routledge.com/9781032515328

Operations Research Series

Series Editors:
Natarajan Gautam
A. Ravi Ravindran

Operations Research Calculations Handbook, Second Edition
Dennis Blumenfeld

Introduction to Linear Optimization and Extensions with MATLAB®
Roy H. Kwon

Multiple Criteria Decision Making in Supply Chain Management
Edited By A. Ravi Ravindran

Multiple Criteria Decision Analysis for Industrial Engineering
Methodology and Applications
Gerald William Evans

Supply Chain Engineering
Models and Applications
A. Ravi Ravindran, Donald P. Warsing, Jr.

Analysis of Queues
Methods and Applications
Natarajan Gautam

Operations Planning
Mixed Integer Optimization Models
Joseph Geunes

Big Data Analytics Using Multiple Criteria Decision-Making Models
Edited By Ramakrishnan Ramanathan, Muthu Mathirajan, A. Ravi Ravindran

Service Systems Engineering and Management
A. Ravi Ravindran, Paul M. Griffin, Vittaldas V. Prabhu

Probability Models in Operations Research
C. Richard Cassady, Joel A. Nachlas

Fundamentals of Stochastic Models
Zhe George Zhang

Supply Chain Engineering
Models and Applications, Second Edition
A. Ravi Ravindran, Donald P. Warsing, Jr., Paul M. Griffin

Solar Power Forecasting
Using Time Series and Machine Learning
Natarajan Gautam

For more information about this series, please visit: https://www.routledge.com/Operations-Research-Series/book-series/CRCOPERRESE

Solar Power Forecasting

Using Time Series and Machine Learning

Natarajan Gautam

CRC Press
Taylor & Francis Group
Boca Raton London New York

CRC Press is an imprint of the
Taylor & Francis Group, an **informa** business

Front cover image: foxbat/Shutterstock

First edition published 2027
by CRC Press
2385 Executive Center Drive, Suite 320, Boca Raton, FL 33431, U.S.A.

and by CRC Press
4 Park Square, Milton Park, Abingdon, Oxon, OX14 4RN

CRC Press is an imprint of Taylor & Francis Group, LLC

Library of Congress Cataloging-in-Publication Data
Names: Gautam, Natarajan author
Title: Solar power forecasting : using time series and machine learning / Natarajan Gautam.
Description: Boca Raton : CRC Press, 2026. | Includes bibliographical references and index.
Identifiers: LCCN 2026008512 (print) | LCCN 2026008513 (ebook) | ISBN 9781032515328 hardback | ISBN 9781032516950 paperback | ISBN 9781003403500 ebook
Subjects: LCSH: Solar energy--Forecasting--Statistical methods | Solar energy--Forecasting--Data processing | Time-series analysis | Stochastic models | Machine learning
Classification: LCC TJ810 .G385 2026 (print) | LCC TJ810 (ebook)
LC record available at https://lccn.loc.gov/2026008512
LC ebook record available at https://lccn.loc.gov/2026008513

ISBN: 978-1-032-51532-8 (hbk)
ISBN: 978-1-032-51695-0 (pbk)
ISBN: 978-1-003-40350-0 (ebk)

DOI: 10.1201/9781003403500

Typeset in NIMBUS ROMAN font
by KnowledgeWorks Global Ltd.

Access the Support Material: Routledge.com/9781032515328

Publisher's note: This book has been prepared from camera-ready copy provided by the authors.

Dedication

To my parents for the inspiration, to my wife for the motivation to write, to my brother for the support, and to my son for all the ideas in shaping this document

Contents

Preface

In the years to come it is anticipated that electricity directly harnessed from solar power would play a significant role in meeting energy demand. However, to achieve that, one of the challenges is that solar power is difficult to predict accurately especially on partly-cloudy days. Further, although storage technology appears to be a way around it, storing unused power requires very large, fast-charging/discharging, and efficient batteries which currently does not exist. Hence, it is critical to be able to effectively forecast solar power to use it effectively.

There are ways to forecast solar power using sophisticated weather models or by modeling cloud formation using satellite imagery. There are also methods based on time-series analysis (as solar power has a nice cyclical pattern) that can be improved using machine learning models. This book is focused on the latter where we consider hybrid models based on both time-series and machine learning. Although there are numerous research articles on this topic, there isn't a book that one could read up to understand the issues, models, and use cases. That motivated the need to create a monograph on the subject.

A good part of this book covers various approaches to make point forecasts. These forecasts can be made in the short term (a few minutes to hours in advance) for which certain methods are presented. They can also be predicted a day ahead (or longer, although that is not considered in this book) using similar methods where seasonality is modeled using time-series methods like exponential smoothing and ARIMA (autoregressive integrated moving average), and then the residuals are used to build an ensemble of decision tree regressors. Thus, anyone who wishes to find a starting point to make such point forecasts can use this book, and then further fine tune using other articles in the literature.

The latter part of this book is about probabilistic forecasting, i.e., forecasting a distribution of solar power supply. When the end goal is to produce a forecast, a distribution is not ideal as one does not know the ground truth. However, when the end goal is to use it in a decision-making-under-uncertainty framework, a distribution is extremely useful. We first present some day-ahead planning decisions and then impress upon the need for distributional forecasts. Then we illustrate approaches of how to obtain distributional forecasts and test them. We also present use cases of micro grids (both stand-alone and partially connected to grid) to illustrate the models as well as describe algorithms.

There are only six chapters in this book and it would be worthwhile to read through all the chapters. That is for multiple reasons. One is that the topics are described only once in the book although they are used in multiple chapters. So it would be useful to read up in the place it was written. Second, there may be some ideas that are mentioned in one part of the book and that may be something you are looking for, although not in the chapter that you were trying to read through. Third reason is that there are python code details that are not provided in all parts of the book, and so it may be important to look through some parts of the book carefully to get those details.

Besides using this book as a stepping stone to understand, appreciate, and use tools for solar power forecasting methods that are in the interface of time-series, machine learning, and decision-making under uncertainty, it can also be used in a classroom setting. In fact, this book came out of a half-semester course on solar power forecasting. It is possible to teach a semester-long course on machine-learning based forecasting using online texts and this book, or expand to wind energy forecasts, or dig deeper into the solar photo-voltaic technology, etc. There are many ways of expanding on this material to be taught as a semester-long course.

Although this book is written with solar forecasting in mind and all the examples provided are related to solar forecasting, it is important to notice that this book can also be viewed as a set of models for short-term forecasts with an application to solar forecasting. So, what is presented is methods that can easily be applied to wind power forecast, sales forecast of each item in a store, demand for number of tickets given a price, number of calls made to a call center, and so on. The models would have to be tweaked but the overall idea still remains the same. This book serves as a use case to learn and apply various prediction techniques for short-term forecasts.

Another important aspect is that we go into fair amount of details about how time-series and machine learning forecasting models work. In fact, that is not presented in many books and it is worthwhile reading through it even if you are not necessarily going to apply solar forecasting. For example, details of how regression trees, gradient boosting, and extreme gradient boosting work is not well explained in many other sources. But there is a lot of emphasis on using them. While selecting the optimal hyper-parameters of the models, it becomes important to understand how they work. So, this book is also useful for someone trying to understand the basics of various machine learning models.

On the day of writing this preface, it is cloudy in Manlius, NY (a suburb of Syracuse, NY). This cloudy day is one of about 200 such days in town in a year. Syracuse is in the top ten cities with the cloudiest days in the United States. But you would be amazed to see how many rooftops have solar installations and how many solar farms are in town. Of course, for a place that is so cloudy, it is extremely crucial to be able to forecast the amount of solar power one would get each day. This way, if needed, one can use the other sources of power. Here in Onondaga County there is a significant source of hydroelectric power. Overall the cost of solar across the state has about halved in 2025 of what it was in 2017. There is a significant improvement and outlook is extremely positive.

Having said that, the hope is that there would be plenty of opportunities to use this book not just in New York but the whole world. We sincerely wish that you enjoy reading this book, and you will find this book useful whether or not you are into solar power forecasting.

About the Author

N. Gautam has been a Professor in the Department of Electrical Engineering and Computer Science at Syracuse University since January 2022. Before that, he was a faculty member at Texas A&M (over 16 years) and at Penn State (8 years). In addition, Dr. Gautam has been an Amazon Scholar since the Fall 2019. His research focuses on optimization and control of stochastic systems with applications in computer-communication networks, renewable energy systems, real-time logistics, and smart manufacturing. He uses probabilistic models, data science, machine learning, and decision-making under uncertainty methods for his work. Dr. Gautam is a Fellow of the Institute for Industrial and Systems Engineers (IISE) and a senior member of both the Institute for Operations Research and Management Sciences (INFORMS) as well as IEEE. He is a department editor for IISE Transactions and, was an associate editor for the INFORMS Journal on Computing. He has written a book "Analysis of Queues: Methods and Applications" and has authored several publications in IEEE, ACM, INFORMS, and IISE journals. Both government and industry have funded his research projects. Dr. Gautam teaches courses in stochastic operations research, analytics, and machine learning. He has received numerous teaching awards. He served as the chief editor of IISE's Industrial Engineering Body of Knowledge, and also was the president of the computer and information systems division of IISE.

1 Introduction

Just to be clear, this is not a book about astrology, i.e., using sun-signs (Aries, Taurus, Gemini, ...) to make forecasts. There actually is a book about that with a similar title, if you are interested, check out Koritnik [25]. This book is also not about trying to predict penetration of solar power into the electricity portfolio across years (Zeng et al [56]). That means trying to forecast across multiple years to predict how solar power would grow. While that is an important problem and a closely related topic, the forecasting techniques used there are different. This book is about making forecasts of how much solar power a panel will produce using historic data from that panel and possibly other exogenous data.

We are interested in forecasting the amount of solar power a panel (or a set of panels) will produce in a short term such as the next 15-minutes until about the end of the next day. So, we have a very near-term analysis throughout. For example, we wish to know how much solar power will be available in the next 15 minutes so that we can decide whether to charge a battery or simply use the solar power. Another example use case is to determine how much of utility power to commit from the grid a day in advance so that it is easy for the grid to make arrangement for that power. For this we need to be able to say how much solar power we would get and buy the remaining power from the grid at a reasonable price.

In all but one chapter, we will mainly be focusing on predicting solar power. There our goal is to make predictions in the short term (such as in the next 15 minutes). In the subsequent chapter we focus on predicting slightly longer term such as a day in advance. Interestingly, we will consider different exogenous features for the two modeling settings. Then in the following chapter, we consider an example of a microgrid which can be thought of as mostly being self-sufficient. There it becomes important to predict not only the solar power but also demand or load as well as electricity costs that the grid will charge us. Finally, in the last chapter, we consider the problem where we need to provide a distribution of solar forecasts.

Although we consider a microgrid, it is extremely important to state that this work can greatly benefit communities that are reliant on solar power. In a recent trip to India, we witnessed a community that is in great need of power for whom solar was an incredibly useful source. The location is near Kabini in the state of Karnataka. Notice from Figure 1.1 where a set of solar panels are used to power an entire household. Notice how the panels, although fixed, are pointing to different directions so they can maximize the amount of power produced. Upon careful viewing one can also see the wires running from the solar panels into the home.

More details of the house is provided in Figure 1.2 which has multiple figurines. The top left and top right are two other solar panels with wires connecting from them into the house. Notice how they are at different angles (compared to the ones seen in Figure 1.1) and within themselves. Also notice that one of them even has a significant amount of shade while the other has some shade. This is an important fact

DOI: 10.1201/9781003403500-1

Figure 1.1 A House in a Remote Village in India Using Solar Power

Figure 1.2 Details in House in Remote Village with Solar Panels

to consider although in many places they are not blocked by low lying trees. But this is indeed an effective way of going around the issue of tree shade, however, nothing can be done with respect to cloud cover.

Next, in the bottom part of Figure 1.2, notice on the left side there is a battery dedicated for the solar power produced. This battery provides electricity during night times and can also be used at times when the power is low. Notice that this is a fairly small-sized battery but is able to cater to the needs of this household. In fact, they even have a nice large music system with amplifiers on the right side of the bottom part of the figure. Notice how small the battery is on the rightmost figure. A man was humming to the music being played. He was having a glorious time. It shows how much a solar panel has contributed to his life.

In a nutshell, whether we live in a well to do neighborhood where we have a choice in power source or we live in a humble remote village with solar as the only source of power, it is critical to forecast and use solar power to the best extent possible. While it may have been obvious to the folks who have access to power only through solar, even the folks living in a well to do neighborhood know that the right thing to do in terms of utilizing a renewable resource is to use solar power. It is our hope that by consuming solar power we can sustain our needs as well as do our part in creating a low carbon environment. The goal of this book is to show the way to do that.

1.1 RENEWABLE ENERGY SOURCES

A renewable energy source is a natural resource that replenishes on a human timescale. That means it can be used continuously without being depleted. Common renewable energy sources include:

- Wind,
- Water or Hydro-power,
- Sunlight or Solar Energy,
- Geo-thermal (Earth's Inner Heat),
- Tidal and Wave Energy, and
- Organic Matter (Biomass).

A wonderful book by Twidell [46] discusses various renewable energy resources and a large fraction of it is devoted to Solar Photovoltaic (PV) power. All the above examples are discussed there.

Unlike fossil fuels, these renewable energy sources are naturally replenished by ongoing processes and are crucial for a sustainable and clean energy future. In fact, fossil fuels such as oil, coal, and natural gas are examples of non-renewable energy sources. Nuclear, is not a fossil fuel but also an example of non-renewable energy source. Although fossil-fuels currently produce over 60% of the electricity in the world, it is expected that this would come down over time. One of the main reasons being that there is only so much fossil fuel available and we will soon use them all. These fossil fuels will not replenish on a human timescale.

However, renewable energy sources, especially solar and wind, have challenges. The amount of wind power and solar power greatly depends on the amount of wind and sunlight. Thus, unlike fossil fuels, renewable energy cannot be created whenever necessary, at the levels needed. The only way that can be done is through energy storage devices such as batteries. Other than that, it also becomes important to be forecasting the amount of wind and solar power. An excellent resource for wind power forecast is Ding [12] that is highly recommended. But this book is for forecasting of solar power. We will discuss these challenges more in the next section.

Thus, despite the challenges, the renewable energy sources are the only type of resources with natural cycles. They are also sources that are abundant and are not depleted when used. The biggest benefit is that they emit fewer greenhouse gases and pollutants compared to fossil fuels. Hence, they are vital to address climate change by reducing carbon emissions and meeting climate goals. By relying on self-replenishing renewable energy sources, we can transition away from fossil fuels. The clean energy, due to the sources being carbon-free leads to healthier environments. Hence there is a strong case to forge ahead with renewable energy.

All the major sources of power, except for solar, use turbines. This includes both renewable sources such as wind, hydro-power, and geothermal energy as well as non-renewable sources such as fossil-fuel and nuclear. Thus solar is somewhat special that way as it uses photovoltaic effect to convert sunlight into electricity without any moving parts. However, it must be clarified that solar is not the only form of power without a turbine. Other examples including electrochemical power sources like batteries and radioisotope thermo-electric generators used in space also do not use turbines. However, clearly the solar PV is the most significant by far.

PV solar panels are used to convert sunlight into electricity. It uses a process called photovoltaic effect where photons from the sun excite electrons in semiconductor material. This allows for the creation of the flow of electrons which results in electric current. This electricity that is generated is direct current (DC) and it is channelled through the panel's wiring, converted to alternating current (AC) by an inverter, and then used. One article that talks about such power electronic issues is Vakacharia et al [47]. But there are many such articles and if one is interested, one could search through the vast body of literature.

While there are numerous articles in power electronics, one paper that is extremely relevant is Blaabjerg et al [7]. The paper talks about using power electronics technology when the renewable energy becomes large scale. The authors show that, by using power electronics for grid integration, it is possible to achieve large-scale renewable energy generation. They discuss technical issues and requirements with special emphasis on wind, solar PV, and energy storage systems. They present research perspectives for the future large-scale renewable generation technologies by incorporating more power electronics systems.

That brings up the topic of production and consumption scales for renewable energy systems. We are going from a utility-scale which refers to a large-scale power generation project (like a solar farm or wind plant) to a utility grid which is the entire network that transmits and distributes electricity from power plants to homes and businesses. So we are talking about wind farms and solar farms with large amounts of power. Our goal is to look for opportunities to aggregate across farms, both of wind and solar. Also, the effective way of doing that is through the use of an energy storage device. That is the eventual goal of this work.

1.2 SOLAR POWER DATA AND CHALLENGES

In Figure 1.3, the solar power produced at a location in Colorado (Boulder) from the National Wind Technology Center is presented (data source Jager and Andreas [23]). It is for the 2012 October days, and what is actually presented is irradiance and not actual solar power. We will discuss irradiance in the next section. Notice how on days such as October 2, 17, and 18 the amount of solar power is a nice curve. That is because on those days there are no clouds. On other days such as October 7 and 8, there is a little bit of clouds at some time causing it to vary slightly from those on October 2, 17, and 18. Then there are days such as October 3 and 22 with a decent amount of clouds.

However, there are some significant overcast days such as October 4, 5, and 6. Also some days that have enough solar activity such as October 12 and 13 on some time of the day and overcast on the others. Then there are days such as October 19, 28, and 29 which are not as overcast as October 5 and 6, but are overcast enough that there is not one moment that it looks like the full solar power. Notice that this is just one month of one year in one location. This location, has more than 200 days of what we call as pure cloudless days (of course October is a rough month). That brings up the most important challenge when it comes to solar power data.

In this book we mainly concentrate on solar power forecasting and our goal is to focus on short term such as 15-minutes ahead to a little less than 2 days in

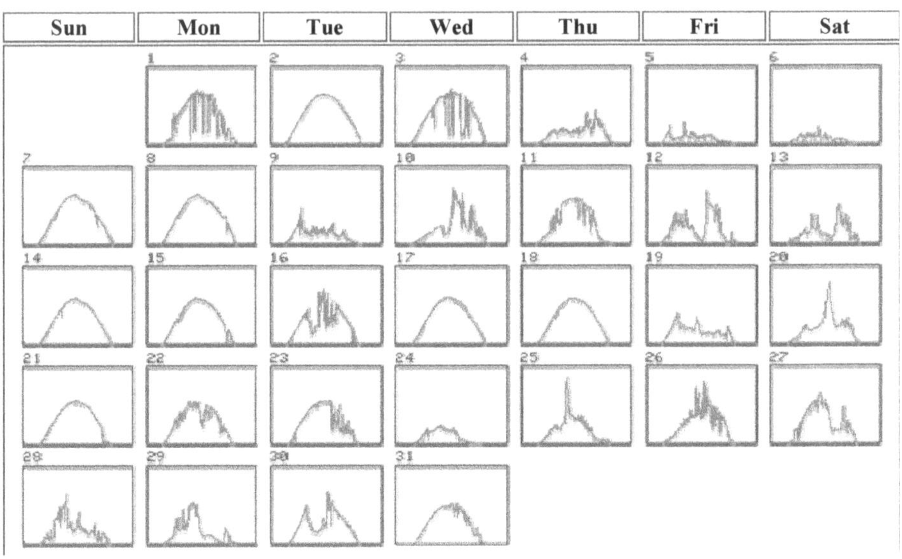

Figure 1.3 Solar Power from a Panel in Boulder Colorado in October 2012

advance. For that, various machine learning methods are extremely useful. Toward the latter part of this book we will look into microgrid (or nanogrid) situations. That is because those are ways by which individuals can operate solar panels for their own consumptions and use it by taking responsibilities for the challenges due to variability. However, nothing prevents a utility provider to own solar panels and predict the total amount of power produced in each time period.

That said, let us briefly describe what is a microgrid and a nanogrid. Details will be provided later in the book. One can think of a microgrid as a campus such as a university, or living community (such as a neighborhood), or a set of office buildings owned by a single entity. A nanogrid is a smaller version of that such as one house or one building. The setting is such that there is one or more solar panels in the microgrid or nanogrid. This is typically behind-the-meter, hence all the energy that is generated by the solar panels are consumed internally and any power that is not used is sold. Also, if extra power is needed, it is bought from a utility.

Typically such microgrids or nanogrids with behind-the-meter arrangements have solar arrays/panels, wind turbines, and batteries. We do not consider wind turbines in our analysis. Since the area is pretty small, the amount of power consumed is also small. Now, we can also consider a slightly larger area where there are multiple solar panels whose inputs are used to drive the demand. Here the demand is larger and so is the solar power. It is fairly atypical to go extremely large scale with a single source but that is also another possibility that we do not consider in this book. Forecasting-wise it does not make a difference though.

Where it does make a difference is that smaller amounts of power implies large amount of variability. This is from a demand standpoint. From a solar supply standpoint, it could mean different things. If panels are very close to one another, then they can be thought of as a single source although partly cloudy conditions do prove to be an issue. If the panels are completely away, for partly cloudy conditions they can be independent but from an overcast or a clear-sky condition, they would still be dependent (of course it does depend on how far they are). Our goal is to take a single panel and say that while adding up the effect of multiple, one has to be careful.

Hence the challenges of microgrids (as well as nanogrids, and to a large extent even the much larger utility scale grid) is the non-stationarity, high variability, and high correlations of the solar supply. That means that the solar supply varies from day to day and hour to hour (this is the essence of non-stationarity). It is also highly varying as a little cloud could change the amount of power significantly. Also, because the weather patterns affect nearby locations greatly, they are highly correlated. So, challenges posed by solar include multiple seasons (daily and yearly), night time no data, and cloud cover dynamics. These are true for all forms of solar energy, not just microgrids.

It is not just power supply but also energy demand (and energy costs which are usually dependent on power supply and demand). So, energy load or demand is also highly *non-stationary*. They depend on the time of the day, and whether the day is a holiday or not. Also, weather conditions would result in high *variability*, and demand can have tremendous variance. In addition, due to spatial proximity, both power supply and power demand would be extremely *correlated*. Further, storage devices have issues of efficiency, storage capacity, and charge/discharge rates. So, while modeling the various parts one has to be very careful.

We ask the question: if the microgrid's average power supply locally exceeds average demand, with a large enough storage device, could a microgrid be completely off grid? If the storage device is efficient enough, it may be possible as long as not too many days are overcast and with very little sun. This is certainly an issue and hence it would be best to have a option where one could connect to the utility grid. For that reason, the examples of solar power are from places like Colorado and California where there is a good number of week days with adequate amount of sunlight. While this might be an issue, it is always possible to have the option of having access to the grid.

In 2013 CAISO (California Independent System Operator) coined the name "duck chart" after they plotted how much net power is produced after subtracting the demand minus supply [51]. The duck curve or duck chart is depicted in Figure 1.4 and is taken from NREL [43], although they credit it to CAISO. The main point is that this curve is plotted against the time of a day on a typical sunny day in spring (May 31, 2013). Notice that in the morning and the evening hours there is a large amount of load demanded. The solar supply is non-zero during the day and is highest around noon (hence lowest net load).

As a result, if one plots the net power (demand minus supply), one would get a curve that looks like that in Figure 1.4. Notice the duck-like shape with the duck

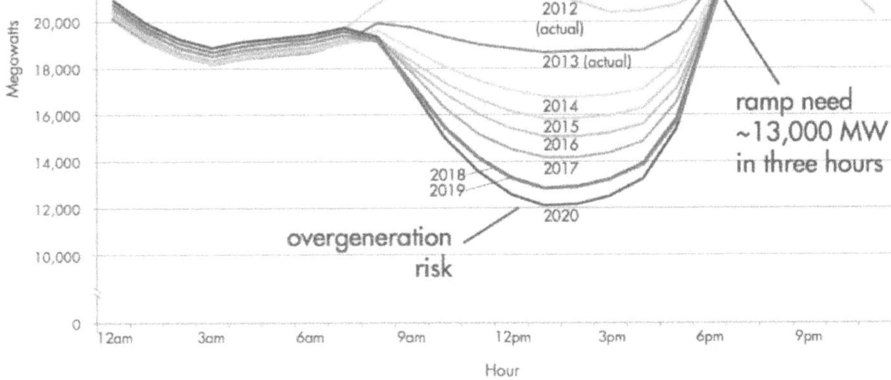

Figure 1.4 Duck Curve of Net Load in California

turning rightward. The tail of the duck is the left side of the figure while the head is to the right. Its body is the shape of the middle part which is mainly causing the variability. Although this is predicted in 2013 based on actual data of 2012 and 2013, as well as forecasted data afterward, other researchers such as Yang et al [55] show that this is still the case with more recent data. So this is another challenge that we explain next.

Note from the duck curve in Figure 1.4 that because of the increase in solar penetration, the amount of supply gotten through solar is increasing but that is available only between 8am and 5pm (roughly). Notice that as the years go by the amount of solar power is so large that the stomach portion of the duck keeps growing. The two dangers are written up on the graph. One is the over generation risk. We are nearly reaching a situation where the amount of solar power may exceed the demand. Hence the risk is over generation. While this can be avoided through the use of batteries, it is still a cause for concern with prices which could go negative (that means to sell power, they would have to pay).

The second problem is that the heavy load occurs in the evening. This is when people return home, turn on devices at home, and warm up their homes. At the same time, this is when the solar power goes away. However, it calls for about 13,000 MW of power needed through other sources. This is a major challenge for utilities. Hence, the duck curve is a graphic representation of daily electricity demand that shows a dip in the middle of the day due to high solar power output and a steep increase in demand in the evening as the sun sets. This creates challenges for grid management, including overproduction of solar energy in the midday and the need for a rapid increase in power generation to meet peak evening demand.

During the midday dip, there can be an oversupply of solar energy, leading to solar curtailment, where power generation is intentionally reduced to prevent grid instability. The steep evening ramp requires utilities to quickly dispatch other power sources, often traditional ones like natural gas, which can strain grid infrastructure and increase costs. Traditional power plants are not always able to quickly ramp up and down to match the fluctuating supply from renewable sources, making grid balancing more difficult. While these are true and tough problems for grid operators they are indeed solvable.

Solutions to the duck curve include battery storage, which flattens the curve by storing solar energy for later use, and flexible power generation sources that can quickly adjust their output. Energy storage systems, particularly large-scale batteries, can store excess solar energy generated during the day and discharge it during the evening ramp, effectively flattening the curve. Generating flexible generation such as developing power plants that can change their output levels rapidly is crucial to meet the steep evening ramp. Demand response by encouraging consumers to shift their energy usage to midday when solar is plentiful can help balance supply and demand. Connecting power grids across larger geographic areas can help balance solar supply and demand over a wider region.

Although it appears like these problems are only for utility scale, they are also the case for microgrids and nanogrids. This is because eventually the net load (whether the solar is consumer behind the meter or not) would still be the same. However, the solutions are surely for utility grids. We do not explicitly address this situation but explain the issue of the amount of power continuously increasing on the power grid in Belgium. Nonetheless, it is important to consider this issue and it is our goal to be cognizant as well as appreciative of the efforts to rectify it. Also it would be good to understand the implications from a cost standpoint.

1.3 BACKGROUND ON SOLAR IRRADIANCE

In the previous section we briefly talked about solar irradiance in Figure 1.3 but then immediately reverted to solar power. Here we take the time to explain irradiance and also see how it is similar to solar power. For that, consider the Figure 1.5 from NREL where we see three lines instead of one line on each day in Figure 1.3. The red (i.e., inner) in both figures are exactly the same, which is the global irradiance. It is also known as global horizontal irradiance (GHI) in Lowry which is in Denver, Colorado. The Boulder location in Figure 1.3 is about 30 minutes of drive north-west of Lowry depicted in Figure 1.5. It is also for the month of October 2012. Notice the similarity in the graphs and that is because they physically close to each other.

However, the green (i.e., outer) and the blue (i.e., lower) in Figure 1.5 need to be explained. The direct normal irradiance (DNI) in green is the solar radiation that travels in a straight line from the sun without being scattered or absorbed. It is measured by pointing a pyrheliometer directly toward the sun. It is harnessed by concentrating solar power technologies. Although not common, it may be possible to receive this much of power by always making the solar PV panels tilt directly toward the sun.

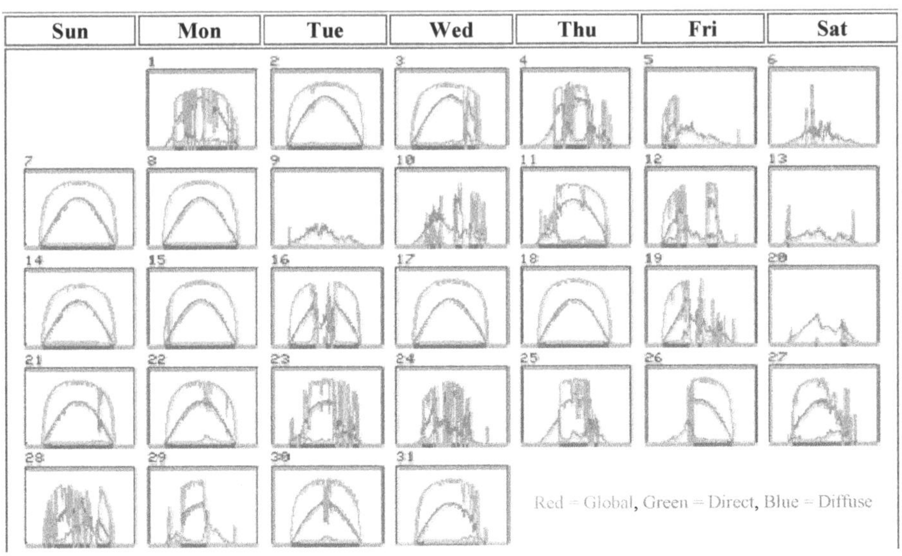

Figure 1.5 Solar Irradiance from a Panel in Lowry Colorado in October 2012

Note that in comparison, the GHI is measured like as though the PV panel is held horizontal on the ground.

The one in blue is the diffuse horizontal irradiance (DHI). It is the solar radiation that has been scattered by the atmosphere such as clouds and aerosols. It arrives from all directions in the sky. It is measured using a pyranometer with a shading disk used to block the direct sunlight. So GHI is the sum of DNI and DHI as measured on a horizontal surface. So, $GHI = DHI + DNI\cos(\alpha_{zenith})$ where α_{zenith} is the sun's zenith angle. A pyranometer with a hemispherical view is used to measure the GHI. But this is fairly common and any static (and horizontal) PV panel would produce that. That is what we would use henceforth.

The solar zenith angle (SZA) is the angle between the sun's rays and the vertical (zenith) line directly above an observer on Earth's surface. It represents how high the sun is in the sky, with a lower SZA meaning the sun is higher and closer to being directly overhead. The SZA varies with the time of day, the day of the year, and the observer's latitude, and it is a complementary angle to the solar elevation angle. The SZA changes based on time of the day, day of the year, and the latitude (the observer's geographical location). A smaller SZA indicates the sun is higher in the sky, while a larger value means it is lower.

It is a common point of confusion that DNI can be higher than GHI at certain times. This is not an error, but a consequence of their definitions. During the morning or afternoon, when the sun is low in the sky, a sensor tilted to face the sun (measuring DNI) receives more concentrated radiation than a sensor lying flat on the ground (measuring GHI). The relationship between them is trigonometric as shown above in the previous paragraph. Since GHI depends on the cosine of the sun's zenith angle, it will be less than DNI when the sun is not directly overhead.

In this book, we mainly focus on GHI for solar irradiance. Although the methods would not change if we considered DNI. Irradiance is the amount of light energy from "one thing" hitting a square meter of "another" each second. Solar irradiance we are interested in is the amount of light energy from the sun hitting a square meter of a PV panel each second. Power (in Watts) is the rate (per second) of energy (in Joules), so irradiance is measured in W/m^2. Irradiance can be measured for any glowing object, including stars, the Moon, and the overly bright high beams of an oncoming car. Human beings radiate primarily infrared light; an infrared image of a human shows a very active heart and mind (from NASA [33]).

Next we discuss power production versus solar irradiance. Power produced by a solar PV panel is a function of irradiance, cell temperature (which is a function of humidity, ambient temperature, pressure, wind factors), and angle of incidence primarily. However clouds, fog, dust, and pollution absorb and scatter sunlight, reducing the solar irradiance that reaches the earth's surface. This can cause the power curve to be lower and more irregular than on a clear day. Also, as a PV panel heats up, its efficiency decreases, which can flatten the peak of the power curve on very hot days. So, there is a difference between solar power and irradiance, and it is not truly a straight line. But it is also important to note that we measure solar irradiance due to its easy availability.

So throughout the book when we say solar power, we do mean irradiance. In most places from the units it would be obvious. However, when we need to focus on the power, we just assume that we can simply multiply the irradiance by the area. That is rather simplistic and does not do justice to the relationship between solar power and solar irradiance. However, we take the commonly held assumption that it is a straight line. It is not a terribly bad assumption especially if we are measuring it for one panel at a time. With that assumption we do say that the irradiance is a straight line relationship with the solar power.

The other main aspect is that the irradiance is almost a fixed value at a given height but the power changes. We call that as power-loss, typically. The power losses can be classified into two categories. The first category is non-avoidable losses that are influenced by factors such as the atmospheric conditions, tilt angle, PV panel cell mismatching, orientation, and PV panel degradation. The second one is avoidable losses caused by dust, soiling, and temporary or permanent shadowing. All these factors significantly affect the performance of PV power generation and are difficult to distinguish. However, from our standpoint we don't make a major difference between the two: solar irradiance and solar power.

1.4 DATA: SOURCES AND FEATURES

In this section we first talk about the data sources and the features used in the modeling. We also explain some of the python related background. After that we give a brief overview of related work in the literature. All the US data have been collected from individual solar panels from various sites (NREL and others maintain this data). A majority of the solar data we will see in this book are from NREL [34]. We also collect solar PV aggregate data from a Belgian site Elia [6]. While these are only the solar PV data, we also collect several weather-related data from Kaggle. These are typically what is used in our work, any other data or their generation would be mentioned in the relevant portion of the book.

The way solar data is presented in the NREL site is that it is measured every minute. So when we have to take the average of 15-minute intervals we simply take the average of 15 of those data points. Likewise for 60-minute data we use each 60-minutes. As a standard practice, we use minutes 0 to 14, 15–29, 30–44, and 45–59 as the 15-minute granularity. Likewise for the 1-hour granularity we use 0–59 minutes of each hour. Technically, in python we simply take the "hour" field and group each hour's data but taking the average across minute 0 to minute 59. Once we have this data, we do not go back to the minute-based data. We also do not compute moving averages.

From a machine learning standpoint, all our models require multiple features. One is the lag of the main predictive variable. So, if we are predicting solar irradiance over 15-minute intervals in the next 15-minutes, we use the current solar-irradiance, a few other sample historical values, possibly also including the same time previous day. In addition, we also use other features such as temperature forecasted, wind speed forecasted, and any other data we can find which are from forecasts. We will explicitly discuss these in our data sets when we study them.

Likewise when we forecast for a day ahead, then we consider lag features of hourly data from a day ago, as well as other samples of lag such as 2–3 hours earlier and also 2–3 days earlier. In addition, we also provide various features corresponding to the forecast of the data such as weather and temperature. In addition, when we consider multiple models, we can build hybrid models by using one's predictions as features of another. Alternatively, we can also build hybrid models by considering residuals. There are many such methods and they would become clear as we develop them in the analysis.

We will be writing our codes for all our analysis using Python. Although there are many ways to write Python files, we will use Jupyter notebook. The main benefit of that is to be able to divide up a code into multiple cells and run each cell. This way we can know where the error is quite easily, if there is an error in some place. However, converting to a generic Python code is also not a problem. Basically, combine all the cells into one big code. In all our Python codes used in this book, our analysis has an initial cell which is given here. That is Cell 1 as shown in Figure 1.6. It is worthwhile explaining what it does next.

The first line with a # in front is essentially commented out. However the # is there to explain what the cell is doing. As it explains, it is the set of commonly used functions in the first cell of a notebook. In the second line (where the number 2 is displayed) essentially states that to run the code we use a python library called pandas. It is extremely useful to generate data frames and there are a lot of programs in Python written with this library. Another library is in line 3 and that is the numpy library. It is extremely useful to obtain linear algebra and other mathematical functions, although other libraries can also be used.

Line 4 essentially states that all the outputs, if float, should be represented as terms rounded off to 2 decimal places. This avoids keeping very long decimal values. However, the actual text does have the full decimal representation. Then, lines 6–8 are specific to IPython/Jupyter Notebook environments and control how outputs are displayed. Line 6 imports the InteractiveShell class, which is the core object managing code execution in IPython/Jupyter. Then normally, in a Jupyter cell, only the last expression's result is displayed automatically.

Typically, the above is for example a print command. Line 7 changes that behavior so that all expressions in a cell will display their output (not just the last one). This way we can avoid writing `print` lines every time we want something to be printed. Then line 8 imports functions to display rich content in Jupyter. For example, `display()` explicitly show objects (e.g., DataFrames, images, plots), and `HTML()` renders a string of HTML as formatted output. Finally lines 10 and 11 essentially imports the warning library and asks Python to ignore warning messages (to the extent possible).

In the next chapter, we will talk about the various error metrics (using numerical examples to illustrate the metrics). Following that, we would describe some very short-term predictions such as the next 15-minutes to an hour out in the next chapter. Then, in the subsequent chapter we discuss forecasts a day ahead. The last two chapters will focus on a case study for making decisions with solar in one chapter

```python
# Commonly used functions typically on 1st cell of notebook
import pandas as pd
import numpy as np
pd.options.display.float_format = '{:,.2f}'.format

from IPython.core.interactiveshell import InteractiveShell
InteractiveShell.ast_node_interactivity = "all"
from IPython.display import display, HTML

import warnings
warnings.filterwarnings("ignore") # specify to ignore warning messages
```

Figure 1.6 First Cell of a Python Code in Jupyter Notebook

and the notion of forecasting distributions in the final chapter. However, in the next few paragraphs we will first explain a little about the literature on solar forecasting depicting the state of the art. Then we will do the same with time series and machine learning methods with an emphasis on forecasting in general.

One of the first articles on solar forecasting methods that got a lot of citations is Inman et al [22]. The authors review the theory behind forecasting methodologies, and a number of successful applications of solar forecasting methods for both the solar resource and the power output of solar plants at the utility scale level. They show that to do that, high accuracy forecast systems are required for multiple time horizons that are associated with regulation, dispatching, scheduling and unit commitment. They take advantage of sky imaging and numerical weather prediction in their results.

Subsequently numerous articles were written in the general topic of solar power forecasting with emphasis on methods such as in Tuohy et al [45]. A more recent paper by Barhmi et al [5] talks about various solar forecasting techniques and describes the role of artificial intelligence. Singla et al. [40] provides a review and analysis of various solar forecasting methods. Yagli et al [54] considers hourly solar forecasting using numerous machine learning models. The article Pellend et al [35] describes the solar forecast as a literature review which is an excellent resource.

An outstanding resource on solar forecasting is the edited book by Kleissl [24]. The book starts with terms and definitions that would be a useful for one to learn more about PV, solar power versus irradiance, and the various forms of solar power radiation. The book heavily uses sky imaging systems and numerical weather prediction as methods for determining the amount of solar power that one could see. There are numerical case studies throughout the book. While that is an extremely good book, it is also important to say that the focus there is more on the solar aspects although our goal in this book is through machine learning.

In a more recent article with numerous references, Yang et al [55] review solar forecasting, how it depends on atmospheric sciences, and the implications for grid integration to get carbon neutrality. It discusses common pitfalls of solar forecasting research with remedies that are proposed. It gives many technical aspects of solar forecasting with a special focus on grid-integration implications. Also, it provides research topics for scientists. It also provides a pathway toward ultra-high PV penetration. Since it is also a review article, it points to more than 250 references in solar forecasting that would be good to review.

Voyant et al [48] reviews a set of machine learning models for solar forecasting. They divide their work into three pieces. They give some methods using artificial neural networks, then talk about tree-based methods, and finally go over support vector machines. Also, Wan et al [49] describes PV and solar power forecasting with an effort to emphasize smart grid energy management. It provides a comprehensive review of the theoretical forecasting methodologies for both solar resource and PV power. Applications of solar forecasting in energy management of smart grid are also investigated in detail.

Forecasting, especially using time-series models, is extremely well explained in Hyndman and Athanasopoulos [20]. It is an online introductory-level textbook that can be freely downloaded. Although it uses time-series models implemented in R, there are Python libraries available in other sites. There are numerous real-life datasets to analyze. Hyndman's website (`https://robjhyndman.com/`) and also his blog are excellent resources. The book (Hyndman and Athanasopoulos [20]) does not do machine learning based forecasts. For that, the course created on Kaggle by Ryan Holrook is an excellent resource to start with and pick up by trying out several exercises there (`https://www.kaggle.com/learn/time-series`).

However, if one's ambition is to become an excellent forecaster, the book to read is Tetlock and Gardner [44]. It is a beautifully written book entitled *Superforecasting: The Art and Science of Prediction*. It is a New York Times bestseller and it is written for a broader audience. It is written from a management and leadership standpoint. However, if not anything, any reader must take away the Ten Commandments for Aspiring Superforecasters. That said, let's jump into our book.

2 Forecasting

In almost every field there is forecasting. For example, if you are running a retail store with multiple items, it become extremely important for you to know how many of each item would be sold on a given day. This would help keep the right number of each item in inventory. In particular, a significantly larger inventory than demand for an item could mean that other items needing more space are bring compromised. Whereas, if the inventory level is too small compared to the demand, then very soon customers would stop arriving at the store. So, the ability to forecast the number of items becomes extremely important. It goes a big way in determining the success of a retail outfit.

Consider airline ticket pricing. For a given source to destination, if the volume is predicted in terms of the number that will accept a price, it would be immensely useful to forecast that information. In particular, if the price is too low, then a number of tickets would be sold off well in advance. While this is good, it does not allow the airline to make the most in terms of profit and thereby would lead it to an eventual fall out. On the other hand, if the airline charged very high prices, chances are that the tickets would not be sold and the flight may have to go without all the tickets booked which is also a waste. So it is critical to be able to price it at a point that would get the necessary volume.

In call centers it becomes extremely important to have a reasonable idea of the demand forecasted. This way one could know how many calls to expect in each hour of the day. Having too many employees at the call center is a problem from a cost standpoint. But having too few would mean that a large number of calls would not be attended to in a timely fashion. Thus there is a trade-off in terms of how many calls one can expect. As more and more companies switch to robot call centers, it becomes even more important to forecast as the number of human operators become much smaller and a careful modeling is necessary as there is more variability.

Consider a healthcare situation where hospitals manage both resources such as doctors and nurses as well as a number of patients. Having a good forecast of the number of patients that arrive in each hour become important. If the actual number of arrivals is too much in comparison to the forecasts, then the patients will not be treated well, leading to possible adverse costs such as insurance liability. On the other hand if the actual number of arrivals are too few compared to the forecasts, then the doctors and nurses are underutilized while they may be desperately needed elsewhere. Hence it becomes important to forecast accurately.

In fact, planning staffing levels is in itself a direct consequence of forecasts. Having a rough idea of the number of customers is extremely useful in determining the staffing levels, whether it is a grocery store, restaurant, movie theatre, package deliveries, or sporting event. Being able to forecast the right number of customers would mean that we can schedule the right number of staff. Having too few has implications

DOI: 10.1201/9781003403500-2

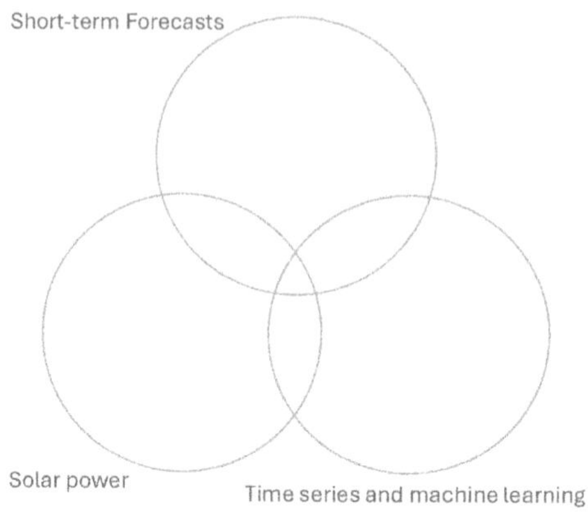

Figure 2.1 Venn Diagram of Various Important Aspects of this Book

on quality while having too many has issues in terms of costs. Thus it becomes extremely important to forecast the number of customers in each hour (or even smaller or larger intervals) each day.

Also, usually from a finance standpoint, equipment and resources are often purchased well before staffing is done which is well before when customers arrive. For example, say that you own a manufacturing company and you have decided to use your own fleet of vehicles to deliver your products to various stores. Then first you have to reserve the fleet of vehicles. At that time you need to have a rough idea of how many items are to be delivered. Then as you draw closer, you have to decide how many drivers to employ for delivery. Now the forecasts need to be closer. Then when the deliveries happen, you need to forecast very close to the actuals.

Although we have not spoken about solar PV power forecasting, it is important to realize that there are many other applications that use forecasting. In fact, a majority of this book can be used for any forecasting. The models, methods, and approaches are not only for solar forecasting. Another aspect is that the models considered here for solar PV forecasting (such as machine learning) can be used for other domains as well. These are depicted in Figure 2.1, although the picture is not drawn to scale. Also, it is important to realize that solar forecasting can be done at multiple levels and we are focusing on short term forecasts of less than or equal to 2 days.

2.1 WHAT, WHEN, WHERE, AND HOW?

Although we presented several forecasting examples before this section, *what* we are interested in this book is solar PV power forecasting. That is the amount of power that is generated by a given PV panel at any arbitrary time. It is usually represented as Watts or kW (kilo-Watts) or MW (mega-Watts). It is the instantaneous power. The term Watt-hour or kWh or MWh is energy that is generated by a PV panel which equals the total amount of power generated multiplied by the time it is generated (assuming both are constant). Now, if the amount of power is continuously changing, then the energy generated in time $(0,t)$ is the integral $\int_0^t P(s)ds$, where $P(s)$ is the power generated in the time instant s.

There is also a notion of solar power aggregated across space. We have seen that before, and it is in fact irradiance. The metric for irradiance is in Watts per meters-squared (or W/m^2). Although ideally we need to assume that the irradiance at a location b is $I(b)$ in units of W/m^2 and the net power over an area a is given by $\int_0^a I(b)db$, we make a simplistic assumption that for a small enough region (i.e., when the area is fairly small) the irradiance is a constant across the entire region. Then the Wattage is simply the irradiance times the area. Also the angle made by the solar PV is assumed to be orthogonal to the sun's rays falling on it. Or else, some adjustment needs to be done, especially if we are interested in direct irradiance.

In summary, there are many aspects to be considered while making solar forecasts. However, we make simplifying assumptions that are reasonable. These include the solar power experienced by a PV panel of size A in units of m^2 is constant anywhere on the square of size A. An approximation is also made across time. But it depends on how forecast is aggregated. The approximation itself is a bit different from that for area of panels. There are two key aspects in forecasting: (a) how is data aggregated; (b) how much in advance is a forecast needed. When the sun is shining, solar power is generated continuously.

However, we typically discretize the power generated for purposes of analysis. For example, many datasets we will see in this book, the raw data is at a 1-minute granularity. There are many ways to mathematically discretize a continuous stream of data into chunks of one minute: (i) take one snapshot at the beginning of each minute; (ii) take a snapshot every 2 seconds and average; (iii) take snapshots according to a Poisson process and average. However, in most of the book, we aggregate over even longer intervals of say 15 minutes or 1 hour. Essentially we just average across the 1-minute data (15 values or 60 values depending on whether it is 15 minutes or 1 hour, respectively).

So when we forecast solar power, what we are forecasting is essentially the discretized quantity. So for example, if we aggregate over 15 minutes, then our prediction is what the 15-minute average would be. Now, this may not be sufficient to understand small fluctuations in load and the objective is assumed to be to get the 15-minute ahead granularity. Hence it is important to understand how (a) data is aggregated, and what its implications are. Next, we ask (b) how much in advance do we wish to predict. Usually this is a multiple of the data aggregation granularity. So if the data is at a 15-minute granularity, we would predict 15, 30, 45, or 60 minutes in advance. Of course, we could also predict 24 hours in advance.

So the answer to the *when* question is that it depends on what we are interested in doing. This is particularly, how long a forecast we need to take and with what horizon. We mainly consider two such models in this book: (1) real-time forecasts where we are interested in the next 15-minutes of solar power generated (approximating it as a constant) and our data is aggregated over 15-minute intervals; (2) day-ahead forecasts where we are interested in hourly forecasts a day ahead and our data is aggregated over 1-hour intervals. Although these two are the most common examples, we will also consider other versions of the setting.

It is also important to notice that these are not necessary at all. For example, we could have considered real-time forecasts every 5 minutes or day ahead predictions every half hour. Or for that matter any other combination. However, what is important is that it should be as short as possible. But not so short that it makes it cumbersome to do all the analysis and predictions, and not so long that it makes the analysis extremely approximate. From that standpoint, it is reasonable to consider 15-minute intervals for real-time forecasts and 1-hour intervals for day ahead in the book.

The question of *where* has a lot of interesting ways to look at. We first consider the locations. We essentially use all the data that we could get. This includes from locations where NREL has collected data in the United States as well as Elia in Belgium. For the NREL data, we essentially forecast for three locations: Aurora CO, Golden CO, and Los Angeles CA. Interestingly, many locations are selected with a large number of sunny days each year. Understandably, that makes a strong case for using solar as compared to a place which gets less than half the days as sunny. But it is important to realize that this work is mainly to say that if we were to forecast well, it can be used everywhere.

Another aspect in solar forecasting that is important are to take measurements during the day. While this is obvious, it does cause a lot of interruptions when we consider no forecasts at night time. So, especially for real-time forecasts when we drop the night time data (as they are all zero), it becomes difficult to predict in the early morning hours. This is a huge challenge that one needs a few data points to get an accurate forecast. Also, although we deliberately drop off points at night, the discontinuity does make the modeling a bit problematic. But we assume that we can do that and continue with the analysis.

Although we briefly described it along with when to make forecasts, an important aspect to consider is the issue of horizon. We need to decide what data is used to make a forecast. Sometimes these horizons are fixed such as one period or one day in advance. So here the forecasting period keeps changing every 15-minutes or an hour. Other times the horizons could be varying. This is especially the case when we need to make day ahead commitment of the amount of power we wish to buy. So we make those decisions at some time (such as 4pm) for the entire next day. So at 4pm, the horizon is 16 hours for 8am, 17 hours for 9am, 18 hours for 10am, and so on until 25 hours for 5pm the next day.

Another import consideration is which month of the year do we make predictions. It is important to note that there is so much variability from month to month even on a sunny cloudless day. That is because of the rotation of the earth around the sun at an axis which also results in the variability of temperatures (such as summer and winter). Further, the tilt and rotation are in elliptical orbits causing more variability. Thus it become tricky to predict, but we try and keep our predictions to a single month or a collection of months that are somewhat similar. Also, it is not that feasible to go from year to year as the solar panels get dirty resulting in varying amount of power.

We said that we want to make day-ahead and real-time forecasts. These use short term forecasts (1 minute, 5 minute, 10 minute, 15 minute, or 1 hour forecasts). The next question is *how* we make the forecasts. While this is the key to pretty much the result of the book, we give some preliminary thoughts here. We use time series models and also explore machine learning methods for forecasting. Historically time-series models have essentially used just the data that is needed for prediction. However, of late with machine learning, we have been able to use other exogenous features for making forecasts. That will be a major change in this forecasting book.

It is important to notice that although we have not collected data that are for exogenous features, it is crucial to note that if one were to design forecasting systems with exogenous features in mind, it could seriously improve the results. For example, we consider weather features such as wind speeds which is generally well forecasted based on the previous day. Then we can use it to determine the weather at a location that is a few minutes away to understand the weather patterns there. However, we only consider one location as exogenous, no matter what direction or speed the wind blows. It is recommended that one considers multiple such measurements whether or not there are solar PV panels.

From a day-ahead standpoint, we consider exogenous features such as weather type. Usually it is easiest to forecast when there are no clouds, and is typically one of the highest. The next easy one is overcast days which are both easy to predict and also the level of solar power is easy to forecast. Although the variability in forecast does cause some changes. Also, it is not clear when the overcast effects would lift off. Finally the cloudy and partly cloudy conditions present most of the challenges as it could lead to a significantly different amount of solar power, depending on the location of the cloud. Notice that is an information that is not typically available in weather forecasts.

From a modeling standpoint, we will first consider time series methods such as exponential smoothing and ARIMA (auto-regressive integrated moving average) models. Then we will also go into decision trees which can be used by creating ensembles such as gradient boosting, random forest, and extreme gradient boosting. We would explain all these methods as we dive into the coming chapters in the book. We also go over some optimization in the book to discuss how to make predictions. Recall we saw the costs in the examples of forecasts of applications other than solar forecasts that the costs for going over could be different from going under. Hence it is not necessary we need to be predicting the mean or median.

We will discuss these aspects in the subsequent sections of this chapter and then move to the future chapters for various models to consider for forecasting. It is important to note that we only explain in one place even though we use them in multiple. So, it would be useful to read through the methods in different places from the book. Also, while some folks may not be interested in the details of the methods, it is highly recommended that we go through them because we need to know why something is not working out the way they should. Also, we will provide Python code examples in all the remaining chapters of the book.

2.2 METRICS FOR ACCURACY AND SUMMARY STATISTICS

The most fundamental forecasting is what is called *point* forecasts. In the next section we will also go into other types of forecasts beyond point forecasts such as interval forecasts and probabilistic forecasts. That said, a point forecast is the forecast type that is easiest to understand and verify. The biggest positive aspect about it is that it can be predicted and checked. The checked values are called actuals. But we also refer to it, especially in the machine learning community, as *ground truth*. So we make forecasts a few minutes or days in advance. Then when the ground truth occurs, to know how good the forecasts are we need to check how close the forecasts were to the actuals.

This is known as *point forecast accuracy* and the purpose of this section is to derive metrics to analyze the accuracy of point forecasts. Oftentimes, we are in a situation where we make multiple forecasts and we obtain a summary statistics of that. Before describing those, we wish to point out an extremely important aspect. Although in the actual system we obtain forecasts into the future, while developing the forecasts, we tend to use historical data. Some of the data we call *train* and the remainder as *test* data. What is important is that nothing in the test data ought to be seen while model building for testing.

In fact, it is preferable that the test data is not even viewed and completely kept hidden from the training data (otherwise what would occur is data leakage). Now, let us focus on the N test data points after the forecasting is done when we have both the forecasted values and the actual values. Say that we make N forecasts $f_1, f_2, \ldots, f_N$ of the N test data points. We have corresponding ground truths or actuals $a_1, a_2, \ldots, a_N$ of the forecasted values. Of course if $a_i = f_i$ for all $i = 1,2,\ldots,N$, it would be ideal. However, that is never going to happen. Hence we need some metrics to understand the differences between the values which we explain next.

It is important to state that we cannot use a metric such as the mean of the difference, i.e., $\sum_{i=1}^{N}(f_i - a_i)$. That is because most of the mean-based forecasting techniques forecast values around the actuals. For example, if $N = 5$ and the actuals are 10, 4, 8, 6, and 12 while forecasts are 1, 8, 2, 10, and 19, it would have $\sum_{i=1}^{N}(f_i - a_i) = 0$. Whereas, a forecast of 9, 3, 7, 7, and 11 would have $\sum_{i=1}^{N}(f_i - a_i) = 3$. But if you look at it, every forecast in the second example is quite close. So the mean itself is not a good metric as we have to have a way to penalize however far they are from the actual. Although in some situations such as obtaining the bias in the forecast, they do use it but we will not.

One of the most popular metric especially used in ordinary least square methods that are common in regression is the sum of squared error (SSE). The SSE is defined as (one would have seen in a standard statistics book)

$$\sum_{i=1}^{N}(f_i - a_i)^2$$

where $(f_i - a_i)$ is thought of as the error and the metric we are after is the sum of squared error (as we are summing up the squared error over all N). In the previous numerical example with $N = 5$, the SSE for the first case is 198 while the second example has an SSE of only 5. So clearly, the second method which was much better is now coming up strong.

Another form of the SSE is called the RMSE (root mean squared error) which is the average of the SSE and then the square root. Hence the RMSE is defined as

$$\sqrt{\frac{1}{N}\sum_{i=1}^{N}(f_i - a_i)^2}$$

where from the formula is clear as the square root of $1/N$ of the SSE. That metric for the two examples are thus 6.293 for the first case and 1 for the second case. However, it is important to note that the SSE and the RMSE are different only by magnitude. Thus if something is higher for a forecast in SSE, it will also be higher for RMSE. That does not change at all. However, while comparing with the next method, RMSE is more appropriate.

That is the alternate method now considered fairly standard in many machine learning libraries. It is called Mean Absolute Error (MAE). Mathematically, the MAE is given by

$$\frac{1}{N}\sum_{i=1}^{N}|f_i - a_i|$$

where it is similar to the RMSE, except instead of squaring the function, we take the absolute value. Here the two examples have the values of 6 for the first set and 1 (the same value as RMSE) for the second set. Although the RMSE and MAE are quite similar here, that is not always the case. Next, we show an example of another set where they are different, and then make some comments.

Now consider forecasts A being 3, 4, 2, 5 and forecasts B being 2, 5, 3, 7. The actual values are 3, 4, 2, 17. The RMSE metric evaluations for forecasts A and B are 6 and 5.074, respectively. The MAE metric evaluations for forecasts A and B are 3 and 3.25, respectively. Thus although forecast A predicted accurate in 3 out of the 4 cases, and it was only slightly worse in the fourth case, it was much worse for the RMSE metric. The reason is that the RMSE metric is much worse for large values whereas the MAE metric treats all values about equally. Hence it does better with forecast A than B,

Technically speaking, the value of f_i that is optimal for the RMSE is the mean while that of MAE is the median. As we know, the medians are not terribly affected by outliers. Thus when there are data points with a significantly large outlier group it may be better to use MAE, although RMSE has been the traditional approach. Further, another important criterion is that the model used to predict might be different from the metric used by the ones in comparing methods. That is something to also keep in mind. Many times, the right approach would be to change the predictive model. For example we may use RMSE to model and use MAE for testing, that may not be ideal and is worth considering.

RELATIVE METRICS

So far we have seen some absolute metrics for the quality of our forecasts. Here, we consider relative metrics. We present four metrics and they are all subtly different. The important thing is to select the metric before selecting the forecasting tool and modifying it. In fact, for that reason we also present this before presenting the methods to do the forecasts. It is quite important to make this judgment early enough, that way one is not changing the metric to suit their forecasting model. In fact, in many companies the metrics are oftentimes known in advance, so this is not much of an issue. Also, in a majority of cases the metrics are relative.

One of the most common relative metric is the Mean Absolute Percentage Error (MAPE). It is

$$\frac{100}{N}\sum_{i=1}^{N}\frac{|f_i - a_i|}{|a_i|}$$

where f_i and a_i are the ith forecasted values of the method and actual values obtained as ground truth among the N data points in the test set. The benefits of this method (compared to the closely related MAE) is that here all the actual values are given the same importance. That means when a_i is small we have to be as careful with the error as when it is big. Another major benefit is that it can be represented as a percentage.

The forecasts A and B in the previous example using MAPE are equal to 0.1765 and 0.5012, respectively. It is also often written as 17.65% and 50.12%. Compared to the MAE and RMSE, where the values were only slightly different (although in opposite ways), here they are tremendously different. Here the method penalizes each value with the same amount of importance, no matter how large or small they are. They can be more appropriate in some situations but not so in others. One main benefit is that solar power changes a lot through the day and giving all of them equal importance can be beneficial. While one could make a case for the opposite.

The main issue with MAPE is if the actuals are zero. That is not much of a concern in solar forecasts as we drop the values that are at night time. Nonetheless, it is important to state the result briefly for the sake of completeness. Widely used approach when actuals can be zero (but not recommended unless forecasts are strictly positive) is symmetric-MAPE or sMAPE which is given by

$$\frac{200}{N}\sum_{i=1}^{N}\frac{|f_i - a_i|}{f_i + a_i}$$

and assuming f_i is non-zero. This avoids the sticky issue of having a zero in the denominator, but like we said before, it is not an issue in solar forecasts.

Another popular metric is the weighted-mean absolute percentage error (WAPE) which is given by

$$\frac{100}{N}\sum_{i=1}^{N} w_i \frac{|f_i - a_i|}{|a_i|}$$

given weights w_i which has physical interpretation. There are two special cases that are worth illustrating. One is $w_i = |a_i|$ for all i which makes it 100 times the MAE. While that is not interesting, notice that by selecting weights equal to the actual values, it gives higher weights to the larger absolute weights. So this is more than anything a mathematical equivalence to make a point of.

But the second one is something we will use $w_i = N|a_i|/\sum_{j=1}^{N}(|a_j|)$ so the weights average to 1. In fact we write that as

$$100\frac{\sum_{i=1}^{N}|f_i - a_i|}{\sum_{i=1}^{N}|a_i|}$$

which we call in this book as mean absolute error ratio (MAER). The reason for that is that it is after all the ratio between the mean absolute error over the mean absolute value of the average actuals. For that reason, in some sense this is exactly like the MAE divided by a constant. The values for forecasts A and B (that we showed earlier in this section) in terms of MAER are equal to 0.4616 and 0.5 (or more technically 46.16% and 50%, respectively).

OTHER SUMMARY STATISTICS

Recall that MAPE is

$$\frac{100}{N}\sum_{i=1}^{N}\frac{|f_i - a_i|}{|a_i|}$$

where f_i and a_i are the forecast and actual values of the N points in the test data. In this section we mainly focus on the MAPE but the same idea can be used in all the other metrics. The problem is that the sample mean may be biased, so we use median and other percentiles. That is the key concept in this approach. So from that standpoint this can also be applied to MAE, MAER, or RMSE. We simply explain for the MAPE as the others work in a very similar fashion.

Say that we arrange the average percentage error (APE) values in increasing order. Note that the APE is $\frac{|f_i - a_i|}{|a_i|}$ or multiplied by 100 to get APE percentage. Then, let $APE_{(j)}$ be the jth smallest among the N APE values. So we have

$$APE_{(1)} = \min_i \frac{100|f_i - a_i|}{|a_i|}$$

and

$$APE_{(N)} = \max_i \frac{100|f_i - a_i|}{|a_i|}$$

as the smallest and largest value.

Also, the median APE is $APE_{(\frac{N+1}{2})}$ if N is odd and the average of $APE_{(N/2)}$ and $APE_{(N/2+1)}$ if N is even. That means a large number of the predictions would have an error smaller than the median APE. Likewise, we could compute any other percentile (such as 75th or 95th referred to as P_{75} or P_{95}) by suitably extrapolating to the 75th or 95th percentile. By displaying the MAPE as well as the median APE, 75th, and

95th, we can see how varying the APE is. Usually the values are quite reasonable for the median but worse for the mean and that is usually because the 95th percentile is extremely high.

Before winding up, is is important to discuss one other item that is popular in machine learning. That is the concept of over-fitting. That happens when the in-sample forecast errors (i.e., using the training data) are low whereas the out-of-sample forecast errors (i.e., using the test data) are high. This typically implies that the data has been over-fit. One needs to do regularization of the parameters. This might not make sense now, but we will explain later. But the main point here is that we compute the accuracy of train/test data by comparing the forecasts against the actuals. The goal is to have the train and test data to have similar accuracy.

2.3 POINT, INTERVAL, AND DISTRIBUTIONAL FORECASTS

Suppose that we wish to forecast tomorrow's solar power at 12pm from your PV system. Let us not worry about how we will go about predicting it as that is the focus of our book, but we ask how do we wish to get the forecast. A very popular metric is the point forecast where we might say that it is expected to be 744 W. Our analysis in the previous section was essentially about how close the 744 W was toward what was actually observed. Suppose that the actual was 800 W, would you say it was good? The APE was $56/800 = 7\%$. You can say relative to another forecast if your was better or not. But what about in an absolute sense?

A lot of researchers essentially state that it would not be a good idea to simply predict the point forecast but an interval must also be specified. Now, if the interval forecast says there is a 95% chance it will be in the range of 700-788 W then the forecast appears to be pretty bad. But if the interval was 644–844 W with 95% probability, then it may be alright. But what prevents us from quoting a really large interval forecast? That is an important question and the interval ought to be the most reasonably short interval. In the true sense, if the interval is very large, then it means that there are very few data points or the variability is pretty high. We shall revisit this.

An interval forecast should provide a range of values with a given probability, such as 95% prediction interval for a specific solar power prediction. In contrast, a distributional forecast describes the probability of all possible future values, offering a full picture of the potential outcomes and not just a single interval. While an interval forecast is a summary of a larger distribution, a distributional forecast is the complete underlying probabilistic model. In other words, if you want the 10%, 20%, ..., 90% probability of solar power at a location, then that can be specified. We will see that there are ways to evaluate it.

An interval forecast is the range of possible values for a specific outcome, such as a solar forecast. It is used to determine a confidence interval within which a future event is likely to occur. The example given earlier that tomorrow's solar power at 12pm will fall between 700 W and 788 W with 95% confidence is such a confidence interval. It is often derived from a larger distribution, summarizing it into a useful range for decision-making. In contrast, a distributional forecast is a full probability distribution that shows the likelihood of every possible future outcome. While the interval forecast is usually parametric, the distributional forecast can be either parametric or non-parametric.

The distributional forecast is used to get a comprehensive understanding of all potential future scenarios and their associated probabilities. For example, a forecasting model might show the probability of having 710 W as 10%, 720 W as 20%, and so on, for all possible values. The underlying principle is that it describes the complete uncertainty around a forecast, rather than just the range where a single outcome is expected. The main difference is scope: an interval forecast is a specific range taken from a distribution, while a distributional forecast is that full distribution, representing all possible outcomes.

In the previous section we spoke about various metrics for point forecasts, such as, RMSE, MAE, and MAPE as well as minor variations of these. For the interval forecasts, the most prominent metric is the Winkler score. Suppose that there is a forecast interval of (l, u) at a $100(1-\alpha)\%$ while the ground truth is y. Then the Winkler score for that data point is

$$W = (u-l) + \begin{cases} \frac{2}{\alpha}(l-y) & \text{if } y < l \\ 0 & \text{if } l \leq y \leq u \\ \frac{2}{\alpha}(y-u) & \text{if } y > u \end{cases}$$

which is the length of the interval $(u-l)$ plus a penalty if the observation is outside the interval.

If the interval is (700, 788) W at a 95% probability and the actual value is 800, then the Winkler score is $W = (788-700) + \frac{2}{0.05}(800-788)$ as $u = 788$, $l = 700$, $\alpha = 0.05$, $y = 800$, and $y > u$. Hence $W = 568$. Now, the other interval which is (644, 844) has a Winkler score of $W = (844-644) = 200$ as $u = 844$, $l = 644$, $\alpha = 0.05$, $y = 800$, and $l < y < u$. Thus picking an arbitrarily high value for the interval is not useful. For example, the interval (200, 1000) W although will result in $l < y < u$, it will result in a Winkler score of 800. Thus in summary this is favoring metrics that are as short as possible and yet not so short that most of the time the value is missed, then the penalty is high.

For distributional forecasts, it is useful to obtain the Continuous Ranked Probability Score (CRPS). Recall that in distributional forecasts there is a single point of ground truth, let us call it y. Also, there is a forecast distribution given by the cumulative distribution function (CDF) $F(x)$ which is the probability that the forecast is smaller than or equal to x. Then, the CRPS is given by

$$\int_{-\infty}^{\infty} (F(x) - \mathbf{1}_{x \geq y})^2 dx$$

for a given value of y and $\mathbf{1}$ is the indicator function. This is ideal when we have the whole forecast distribution as opposed to given quantiles or intervals.

We will however be largely interested in quantiles in distributional forecasts. That means we would be interested in a large set of quantiles for which we obtain the CDF. For example, we might be interested in deciles such as 0.1, 0.2, ..., 0.9 as the

CDF or quantiles. Alternatively we may be interested in quartiles, i.e., 0.25, 0.5, and 0.75 as the values of CDF. One can also get all the percentiles from 1% to 99%. Any number of values is possible, however they need to be countable. We have provided a description of such a quantile score or pinball loss function in Section 6.2.

This brings us to a point where we are ready to make forecasts of solar power in PV cells. The next 3 chapters are going to be on point forecasts. Finally in the last chapter we will go into quantile forecasts. Several of the metrics discussed here will be dealt with in detail in those chapters. We sincerely hope that as a reader you will get all the details of solar forecasts.

3 Short-Term Solar Forecasts

The aim of this chapter is to provide a few methods to obtain *short-term* forecasts of solar irradiance. Before forging ahead with the details, we first make some clarifications about the following words in the previous sentence: methods, short-term, and irradiance. It is crucial to understand that there is no notion of *best method*. The effectiveness of the various methods could depend on the location and time-of-the year. The reader is encouraged to try all methods to see which one works best in their situation. In the last section of this chapter, we provide a comparison of the methods, however. Also, it is important to understand the inner workings of the methods to get a full appreciation of the techniques therein.

In this manuscript, by *short term* we typically mean within the next hour or two, for sure within a day. In the next chapter, we will look at day-ahead forecast. However, there are a few related definitions that need clarification. The data is typically represented as a time series with a granularity. In the data sets we use in this book, the irradiance data is at a 1-minute granularity. So, for instance if the data are available from 11am on a particular day, then the rows would correspond to 11:00, 11:01, 11:02, etc. However, based on the application, an appropriate level of granularity may be needed. In this book, we consider 15 minute and 1 hour granularity by averaging over non-overlapping intervals of 15 minutes and 1 hour, respectively.

Say the time right now is 11am and we wish to get forecasts for the solar power averaged over the next 15 minutes, 15–30 minutes, 30–45 minutes, and 45–60 minutes from now. Then we would consider a 15-minute granularity data by averaging over 15-minute intervals and predict 1, 2, 3, and 4 future periods, respectively. We say that the *horizon* is 4. Alternatively, if we desire to make predictions averaged over 1-hour intervals, then we would predict a horizon of 1 step to get the forecast for the average solar power between 11:00 and 11:59. We typically do not use aggregations larger than an hour. So a short-term prediction for 4 hours from now would be made by averaging over a 1-hour interval in the past for a forecast to be made in a 1-hour period that ends 4 hours from now.

Having made a disclaimer about methods, and clarifying the notion of short term, we now focus our attention on *solar irradiance*. We assume that we have a time-series data averaged over appropriate intervals of the solar irradiance (measured in Watts per meter-squared or W/m^2). So the resulting time series data might look like Table 3.1 which is after averaging over 15-minute intervals. Sometimes the "time" column would just say 11:00, 11:15, 11:30, ..., to indicate the 15-minute beginning times. Now, the last column of the dataset is the average irradiance computed by averaging across 15 data points taken 1-minute apart. If the dataset has power generated instead of irradiance, it could have been used instead.

DOI: 10.1201/9781003403500-3

Table 3.1
15-minute average solar irradiance dataset sample

Index	Date	Time	Avg. Irradiance
0	Apr-5-2023	11:00-11:14	632
1	Apr-5-2023	11:15-11:29	641
2	Apr-5-2023	11:30-11:44	579
⋮	⋮	⋮	⋮
T	Apr-12-2023	11:15-11:29	645

Next, let us take a graphical view of the data to understand various nuances. The time series data is from NREL [34]. We use Global Horizontal Irradiance (measured in W/m^2) as the variable that we wish to predict (that website also provides direct and diffuse irradiance, as described in the first chapter of this book). From the website we use data from SolarTAC (Solar Technology Acceleration Center). SolarTAC is in the city of Aurora (suburb outside eastern Denver) in the state of Colorado in the United States. Credit for the data source is Andreas and Wilcox [3]. Denver is known to be one of the sunniest places in the United States enjoying more than 250 days of sunshine in a median year. For that we wish to look at a week's worth of global horizontal solar irradiance data in Figure 3.1.

The data depicted in Figure 3.1 is for a week in 2016 from September 6 to 13. The minute-level granularity of irradiance data from the NREL website [34] is averaged over 15 minutes to produce the graph. The flat lines at the bottom of the graphs correspond to night time when there is no irradiance, also notice that the date changes around midnight. During the course of each day the irradiance increases until around midday when the sun is overhead. Then the irradiance reduces. The jagged lines are

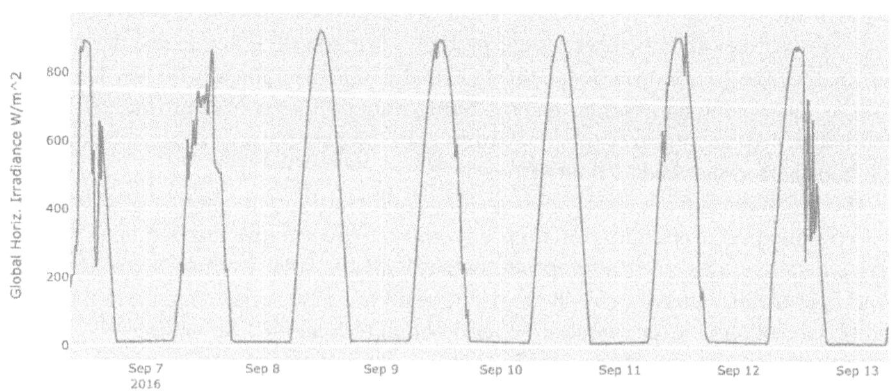

Figure 3.1 15-minute averages of solar irradiance across a week in Sept. 2016

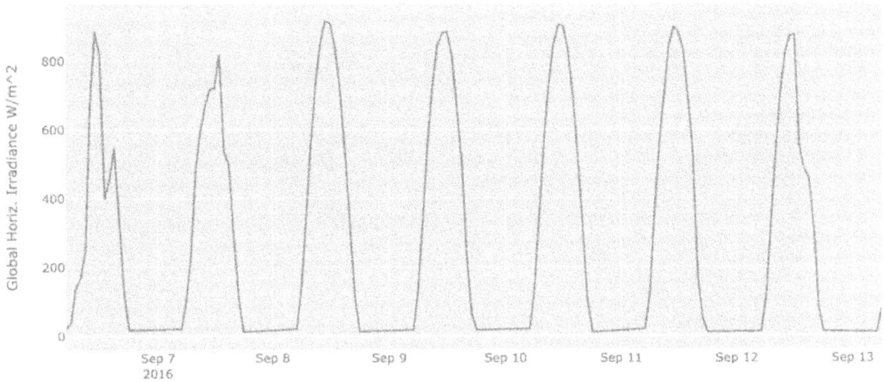

Figure 3.2 Hourly averages of solar irradiance across a week in Sept. 2016

due to cloud formation that blocks sunlight and reduces the irradiance. Notice that in 2 of the 7 days, the curve is absolutely smooth indicating that there were no clouds blocking sunlight.

Now let us turn our attention to Figure 3.2 which spans the same time frame of 7 days as in Figure 3.1. However, the key difference is that in Figure 3.1 the irradiance data is averaged over 15 minutes, while it is averaged over 1 hour in Figure 3.2. Hence the graph in Figure 3.2 is a lot smoother than that in Figure 3.1. Notice that the days that had only minor effects due to clouds get smoothed out greatly. Hence it appears like there are 4 out of the 7 days that are completely sunny. But upon close observation, one can see small variations. When we develop the methods for forecasting, at a certain hour, all points to the left of that hour in the graph may be used to predict.

A longer snapshot of the graph is in Figure 3.3 where the global horizontal irradiance is plotted over a one-month period. At this viewpoint, except for a handful of

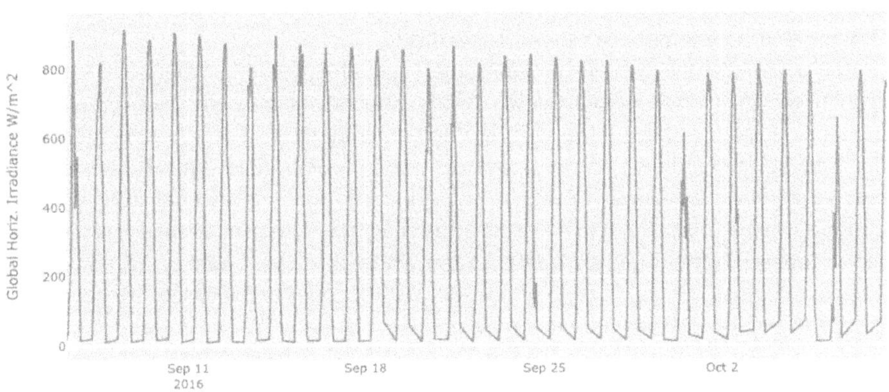

Figure 3.3 Hourly averages of solar irradiance across a month in 2016

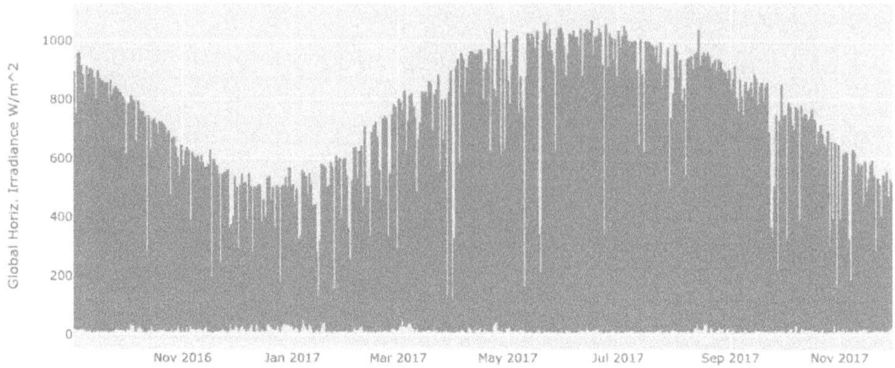

Figure 3.4 Hourly averages of solar irradiance across a year in 2016-17

days the graph looks somewhat smooth. That is because clouds are more significant on those handful of days but the other days were either sunny or had very temporary effects that are not clear. However, there are two aspects to notice. One is that the bottom part of the graphs do not all have flat portions, and that can be attributed to the days getting shorter and hence an hour's average might be small if only parts of that hour has sunlight. The second aspect which is more important is that the peaks each day seem to gradually go down.

In fact, the downward trending peak is because of the earth's rotation around the sun (much like the daily patterns which are due to the earth rotating around its own axis). If we were to look at more than a year's worth of data (see Figure 3.4), then a sinusoidal pattern can be observed as the peaks. It also appears like that trend continues year after year. Note that in Figures 3.2, 3.3, and 3.4, all the data points are averaged across 1-hour intervals. Based on the figures, we can guess that if we were to forecast the amount of solar irradiance that would be generated in an hour, on average, then we would have to know the date, the time of day, as well as some information about cloud conditions that could be gleaned from the previous hour's data and that hour's data over the last few days.

Hence time-series models are extremely suitable for this where we consider trend (long-term effects), seasonality (daily effects), and recent data (recency effects like cloud formation). We will first demonstrate the use of two extremely popular time-series methods, namely, exponential smoothing and Auto-Regressive Integrated Moving Average (ARIMA). Subsequently, we will also consider machine learning models that can also be used in forecasting. Time-series analysis methods like exponential smoothing and ARIMA that we will see are based on Hyndman and Athanasopoulos [20] which is an online text with *R* codes and datasets. To get a good understanding of forecasting, one could go over that book like a course.

We will use the notation in Hyndman and Athanasopoulos [20], that way if the reader wishes to dig deeper, they could easily do so. The *R* codes in Hyndman and Athanasopoulos [20] have also been developed into Python (a software that we will use in this book). For the two methods: *exponential smoothing* and *ARIMA* (with some modifications called SARIMAX that is not explained in Hyndman and Athanasopoulos [20]), we describe some notation. Refer to Table 3.1. We use the letter y to denote the metric we wish to forecast. In Table 3.1, y is "Avg. Irradiance". In particular, we denote y_t as the average irradiance for index t which in Table 3.1 corresponds to the tth 15-minute interval. However, the index could denote other duration lengths.

From Table 3.1, notice that we have indices going from 0 to T which are arranged chronologically. Also, $y_0 = 632$, $y_1 = 641$, $y_2 = 579$, ..., $y_T = 645$. All this information is known and we wish to forecast, to start with, y_{T+1}. Our *estimate* or *forecast* for time $T+1$ given we have observed vales from index 0 to T is represented as $\hat{y}_{T+1|T}$. The "hat" on the y denotes it is an estimate and not what was actually observed. Also, the "|" sign means "given" and what follows the "|" sign is what is known or given. Sometimes we get sloppy and just write $\hat{y}_{T+h}$ for some $h \geq 0$. Then it means that we have observed until time T and wish to forecast h steps ahead (this is also called h-step ahead forecast).

Remark

Recall from the graphs the flat lines at the bottom of the graphs denoting zero irradiance due to night time. For short-term forecasting this is problematic in many ways. First, the methods where we use multiplicative approaches (where we use a "fraction" of the previous observation) would not work. Second, zero values do not carry any information that is useful. So in our datasets we will simply remove the data points at night time. For example, if night time is 6pm to 8am and a 15-minute forecast is needed, then after removing the night values, from 5:45pm, the next point would be 8am. So for predicting 10:30am, we would use observation from 10:15am but for 8am we would use an observation from the previous evening. This would not be an issue in day-ahead forecasting that we would see in the next chapter. However, here we would have to be cognizant of that.

With that caveat, we now present exponential smoothing first, followed by SARIMAX, and ensemble of trees for forecasting.

3.1 EXPONENTIAL SMOOTHING

One of the most fundamental time-series forecasting techniques is Exponential Smoothing (ES). The details of the method are extremely well explained in Hyndman and Athanasopoulos [20] and like we said earlier, we will use the same notation so the reader can look into details in there. Our goal is to briefly go over ES and also adapt it for short-term solar forecasting here (and in day-ahead prediction in the next chapter). The model we will explain next is called *Holt-Winters method with seasonality*. A special case of the model is what is often referred to as ES or even SES (simple exponential smoothing) and typically the first model shown in any time-series course. We will first explain the Holt-Winters method and then say how the special case of ES is derived.

Based on what we saw in Figures 3.1 to 3.4, the key elements that would be useful to make a prediction are: the cloud conditions, the time of the day, and the day of the year. So to make a short term prediction, it would be good to know the current cloud conditions, and the irradiance for the time of the day and year. A reasonable signal to use for current cloud conditions is the most recent irradiance measurements. On top of that we can also use the measurements from the same time the previous day or few days. Also, we can use the trend observed during the time of the year (following the sinusoidal-like curve) in Figure 3.4. In time-series parlance these are referred to as level (for current cloud conditions), seasonality (for time of day effects), and trend (for yearly sinusoidal pattern).

Hence the Holt-Winters' seasonal method with multiplicative seasonality will be a set of recursive equations that are fitted using historical data and make predictions into the future. Recall that we have observations $y_0, y_1, \ldots, y_T$. We wish to estimate y_{T+h} which is written as $\hat{y}_{T+h|T}$. In other words, we are making an h-step ahead forecast. We will oftentimes use $h = 1$ meaning the very next time period. Also, we tend to use a moving window. So first we predict y_{T+h} using 0 to T, then y_{T+1+h} with 1 to $T+1$, followed by predicting y_{T+2+h} using 2 to $T+2$, and so on. Although it is not necessary, we assume that we will train the model for each of the time windows. Hence, without loss of generality we just consider time 0 to T.

The model requires us to also provide the period of seasonality, m. If we collect hourly data and we do not drop any data points, then $m = 24$. In other words to predict at 11am, it would be good to know the irradiance yesterday at 11am. However, note that we drop some data points during night time. If we drop 8 hours, then $m = 16$. Also, if we use 15-minute intervals (instead of 1 hour as said earlier), then without dropping, we would have $m = 96$ and after dropping 8 hours, we would get $m = 64$. One must be careful to pick the right period of seasonality m and the easiest way is to see which data-point corresponds to the exact same time a day ago. Just to clarify, if we were using daily data and there is day-of-the-week seasonality (not for solar forecasting but another application), then m would be 7.

Now we describe some notation. Although y_t is the irradiance at time t, as described earlier, we believe it depends on the cloud conditions (called level and denoted by ℓ_t), day of the year (called trend and denoted through b_t), and time of the day (called seasonality and described using s_t). More precisely, ℓ_t is the irradiance that can be attributed purely to cloud conditions at time t. Clearly, as we saw the sinusoidal variability over the year, so for each unit of time a quantity (possibly negative) is added to ℓ_t and this quantity is called b_t that captures the trend. Also, we can think of a factor that is used to multiply the raw irradiance to account for the time of day. That multiplying factor due to seasonality at time t is s_t. The Holt-Winters' method essentially is a statistical approach to divide up the signal y_t into components ℓ_t, b_t, and s_t.

The recursive equations governing the Holt-Winters' *multiplicative* seasonality model are given as follows. For all t from 1 to T,

$$\begin{aligned}
\text{Forecast equation: } \hat{y}_{t+h|t} &= (\ell_t + hb_t)s_{t+h-m(k+1)} && (3.1)\\
\text{Level equation: } \ell_t &= \alpha \frac{y_t}{s_{t-m}} + (1-\alpha)(\ell_{t-1} + b_{t-1}) && (3.2)\\
\text{Trend equation: } b_t &= \beta^*(\ell_t - \ell_{t-1}) + (1-\beta^*)b_{t-1} && (3.3)\\
\text{Seasonal equation: } s_t &= \gamma \frac{y_t}{\ell_{t-1} + b_{t-1}} + (1-\gamma)s_{t-m} && (3.4)
\end{aligned}$$

where $k = \lfloor (h-1)/m \rfloor$. Before we explain the above equations, it is important to realize what are unknown. The unknowns are α, β^*, γ, s_0, s_{-1}, ..., s_{-m+1}, b_0 and ℓ_0. All the other terms can be written in terms of the unknowns α, β^*, γ, s_0, s_{-1} ..., s_{-m+1}, b_0 and ℓ_0 as well as y_t for all t from 1 to T.

In particular, the unknowns can be used to recursively get the seasonality term s_t, the trend term b_t, and the level term ℓ_t, for all t starting with 1 to T and thereby obtain $\hat{y}_{T+h|T}$. To explain further, notice that when we know γ, ℓ_0, b_0, and s_{-m+1}, using Equation (3.4) and y_1, we can compute s_1 (letting $t = 1$). Likewise ℓ_1 can be obtained using α, ℓ_0, b_0, and s_{-m+1} through Equation (3.2). Then b_1 can be computed β^*, ℓ_1, ℓ_0, and b_0 using Equation (3.3). Next, we can let $t = 2$ and obtain s_2, ℓ_2, and b_2 in that order using the corresponding equations as all the values in the right are known (or in terms of the above few unknowns).

So what is left is to compute the unknowns α, β^*, γ, s_0, s_{-1} ..., s_{-m+1}, b_0 and ℓ_0. The unknowns are computed so that

$$\sum_{t=1}^{T-h} \{\hat{y}_{t+h|t} - y_{t+h}\}^2$$

i.e., the sum of squared error, is minimized. Inside the sample, the model makes a prediction $\hat{y}_{t+h|t}$ but we also know the actuals y_{t+h} for all t from 1 to $T-h$. Software packages can typically compute the minimum point quite fast. The idea is to take the partial derivative with respect to each of the unknowns, make an initial guess and recursively solve for them (interested readers can look up *gradient descent* and its improvements to learn more about how this is done). Finally, once we get ℓ_T, b_T, and $s_{T+h-m(k+1)}$, we can use Equation (3.1) to obtain the forecast $\hat{y}_{T+h|T}$.

In the next few paragraphs we are going to explain Equations (3.1)–(3.4). The forecast equation (i.e., Equation (3.1)) essentially states that if we have data until t, then we would forecast the irradiance at $t+h$ as the level forecast at time t given by ℓ_t as the best guest for h periods ahead, plus any increase or decrease due to trend (for each period it is b_t, hence hb_t in total), and the trend adjusted prediction is scaled by the seasonal term $s_{t+h-m(k+1)}$. The last term may need an explanation. For ease of explanation, say we are aggregating over an hour. It is 11am right now and we wish to make a prediction $h = 3$ hours from now, i.e., 2pm. Then we wish to know the factor used yesterday at 2pm and this is captured by $s_{t+h-m(k+1)}$ with $k = 0$ and hence exactly 24 hours prior to 2pm today. The reason we have the $m(k+1)$ term is because if we want to predict say 27 hours ahead, then $k = 1$ and hence we need to go to 2pm two days prior to when the prediction is needed.

Equations (3.3)–(3.4) are essentially similar and of the generic form

$$\hat{z}_{t+1|t} = \lambda z_t + (1-\lambda)\hat{z}_{t|t-1} \tag{3.5}$$

for some time series $\{z_t, z_{t-1}, z_{t-2}, \ldots\}$ where $\hat{z}_{t|t-1}$ is the forecast of z_t when the time series until $t-1$ is observed, and $\hat{z}_{t+1|t}$ the forecast at $t+1$ when observations until t are made. That result is obtained using the key idea which is to assume an exponential smoothing method that says

$$\hat{z}_{t+1|t} = \lambda z_t + \lambda(1-\lambda)z_{t-1} + \lambda(1-\lambda)^2 z_{t-2} + \ldots$$

where $\lambda \in [0,1]$ is called the smoothing parameter. Note that larger weights are given to most recent observations, and weights attached to the observations decrease exponentially as we go back in time, hence the name exponential smoothing. Since the above is true for any t, we also have

$$\hat{z}_{t|t-1} = \lambda z_{t-1} + \lambda(1-\lambda)z_{t-2} + \lambda(1-\lambda)^2 z_{t-3} + \ldots$$

and multiplying the above by $(1-\lambda)$, we get

$$\hat{z}_{t+1|t} = \lambda z_t + \lambda(1-\lambda)z_{t-1} + \lambda(1-\lambda)^2 z_{t-2} + \ldots = \lambda z_t + (1-\lambda)\hat{z}_{t|t-1}$$

which is precisely Equation (3.5).

The way to interpret Equation (3.5) is that the forecast at time $t+1$ equals a weighted combination of the actual value at time t and the forecasted value at time t with weights λ and $1-\lambda$. This is how Equations (3.3)–(3.4) are obtained. In Equation (3.3), ℓ_t is actually the forecast of the level at time t, and the actual value is nothing but y_t scaled by s_{t-m} which would be the season adjustment. Also, the forecast equals the forecast at time $t-1$ given by ℓ_{t-1} plus the trend adjustment b_{t-1}. The smoothing factor is α.

Likewise, the trend equation for the trend forecast at time t given by b_t in Equation (3.3) is the combination of the actual trend at time t which is the difference between the levels and the previous forecast which is b_{t-1}. The smoothing factor is β^*. Finally, the seasonal equation for s_t the actual seasonal value is attributed as the ratio of the actual irradiance y_t divided by the predicted level at time $t-1$ in Equation (3.4). Also, the forecasted value is that of a period ago, i.e., s_{t-m}. The smoothing factor is γ.

It is worthwhile pointing out that if one were interested in an *additive* seasonality model, instead of multiplicative, then the corresponding equations become as follows:

$$\begin{aligned}
\text{Forecast equation: } \hat{y}_{t+h|t} &= \ell_t + hb_t + s_{t+h-m(k+1)} \\
\text{Level equation: } \ell_t &= \alpha(y_t - s_{t-m}) + (1-\alpha)(\ell_{t-1} + b_{t-1}) \\
\text{Trend equation: } b_t &= \beta^*(\ell_t - \ell_{t-1}) + (1-\beta^*)b_{t-1} \\
\text{Seasonal equation: } s_t &= \gamma(y_t - \ell_{t-1} - b_{t-1}) + (1-\gamma)s_{t-m}
\end{aligned}$$

which can be considered, although the solar forecasting we think of a fraction of solar irradiance that reduces due to the sun's angle. Hence multiplicative appears to be more ideal. However, there are numerical stabilities and ability to provide intervals for the additive, hence it is also presented.

For sake of closure, we turn our attention to the simple ES or SES is where there is no trend or seasonality. Although we do not do this for solar forecasting here, one could scale the solar irradiance data over time to adjust for seasonality and trend. Thus one can potentially consider simple ES. If z_t is the stationary trend and seasonality adjusted scaled irradiance (usually scaled between 0 and 1), then we can write the generic form (as described in Equation (3.5)) as

$$\hat{z}_{t+1|t} = \lambda z_t + (1-\lambda)\hat{z}_{t|t-1}$$

for any t and smoothing factor λ. The only unknowns are λ and z_0 which can be obtained by minimizing the sum of squared error.

Having discussed all those possible alternatives, here we stick to the *Holt-Winters' multiplicative seasonality* model. There are several software programs that can be used. Software also allows for a Box-Cox transformation as well as dampening. For solar forecasting neither is required. Usually Box-Cox transformation is crucial if the data needs to be carefully scaled, and damping is needed if the trends may be too sensitive. Neither is applicable for solar forecasting.

There are few other issues to discuss before delving into the code for forecasting. One is what to choose as T. A frequently used rule of thumb is about 200 points to train as T. For 15-minute granularity data taken for 10 hours in a day, the training is about 5 days. However, the training period T would be about 20 days for 1-hour granularity. In our code, we first only provide point forecasts. That means we provide the expected value of what the forecast would be. After going through the point forecast code, we will discuss obtaining a forecast interval.

To obtain a point forecast of solar irradiance, we use a Python code and provide the cells of a Jupyter notebook. The first cell is identical to that in Figure 1.6. In Figure 3.5 we provide the second cell (represented as `In [2]:` in Figure 3.5). There are four lines in that cell marked from 1 to 4. Line 1 is simply a comment that says what the cell does. In line 2 we import the exponential smoothing model classes and functions from the python module `statsmodels` and in particular under time series analysis (tsa). The entire Holt-winters method explained earlier is coded in the back end of the function.

We also import two functions in lines 3 and 4 of the second cell shown in Figure 3.5. They are from the module for metrics from the sklearn (stands for *sci-kit learn*) and as the name suggests, used to compute the mean absolute percentage error and mean absolute error respectively in lines 3 and 4. We will not use this at present for exponential smoothing but will invoke it when we compare methods. Next we move to the third cell (represented as `In [3]:` in Figure 3.5). There we use `pandas` written in short form as pd (see Figure 1.6) to read from a CSV file. The file is read into a dataframe `input_data` in line 2 of the third cell. To display the header of the data frame, we write down line 3.

Notice the output of the third cell (represented as `Out [3]:` in Figure 3.5). The leftmost column without a name but with numbers 0, 1, 2, 3, and 4 is the index of the dataframe. Then come the two named columns of the dataframe: datetime and global. The datetime column gives the date and time (in format YYYY-MM-dd hh:mm:ss, where YYYY denotes the four-digit year, MM denotes month in 2 digits, dd stands for day, then a space followed by the hours, minutes, and seconds). Notice that the minutes go in steps of 15, that is because the data is averaged over 15-minute windows. The column called "global" is the solar irradiance in W/m^2 that is averaged over a 15 minute interval.

Next we move to the main Python code in the Jupyter notebook cell 4 which (represented as `In [4]:`) is the entire Figure 3.6. The lines that begin with the hashtag symbol are comments that essentially explain what is coming. Hence those lines are not presented and our focus is on what the non-commented lines indicate. In line 2,

```
In [2]:  # Key imports for this code
         from statsmodels.tsa.api import ExponentialSmoothing
         from sklearn.metrics import mean_absolute_percentage_error
         from sklearn.metrics import mean_absolute_error
```

```
In [3]:  # Obtain solar irradiance data as a data frame
         input_data = pd.read_csv('Oct2016_no_night_solarTAC.csv', parse_dates = ['datetime'])
         input_data.head()
```

Out[3]:

	datetime	global
0	2016-10-01 08:00:00	392.13
1	2016-10-01 08:15:00	439.06
2	2016-10-01 08:30:00	483.33
3	2016-10-01 08:45:00	525.62
4	2016-10-01 09:00:00	566.48

Figure 3.5 Cells 2 and 3 of the Python Code for Exponential Smoothing

we create an array called `time_series` by taking the column "global" from the input data frame (called `input_data`). The reason for this is that the time-series needs to be presented as an array (and not a data frame element) when passed on to the exponential smoothing function.

Line 5 `seasonal_periods` denotes the *period of seasonality* that we denoted as m earlier. This is 32 because each day has 8 hours of data (where there is significant sunshine) and each hour has 4 data-points as the granularity is 15 minute average. Recall from the CSV file read in Figure 3.5 that the CSV file only has daytime data and low solar power or dark times have been removed. So, every 32 datapoints denotes the same time but a day apart. In line 8 we say that we make an h-step-ahead forecast for $h = 1$. Then in line 11 we use a 7 day history to train a model and make a prediction h days after.

The code itself has been written for $h = 1$, and no further modifications will be needed for $h > 1$. We will explain the implications of $h > 1$ later as needed. Lines 14 and 15 are empty lists that would be populated with the percentage error and the absolute error made for each forecast at $h = 1$ steps ahead. We will populate the lists in lines 31 and 32. Now we focus on line 18 for the case $h = 1$. This is the part where we do a moving window. We first train using 0 to $T-1$ and predict the irradiance for observation T. Then we move the window by 1 and train using 1 to T to predict $T+1$. Next, we use 2 to $T+1$ to predict $T+2$. We keep doing this until we run out of elements in the time series array.

Thus in line 18, we start with index 0 and go until the last possible index which is the length of the time series minus $T+1$ so that each line in the "for loop" selects a training data of length T. Note that in Python the indexing is from 0 to $T-1$ (as opposed to 1 to T). Also the function `range()` lists all the numbers from 0 until the number in the argument minus 1. So `range(5)` would be 0, 1, 2, 3, 4. Hence the "for loop" in line 18 essentially says that i starts from 0 and goes all the way until the last possible index that would leave at least 1 data point in the set that we can compare against while making the prediction.

In other words, i is the first index in the set of indices of length T that is used to train the data. This is depicted in line 20. The series `train` in line 20 is the array corresponding to elements indexed from i to $i+T-1$ (this is written as $[i : i+T]$). Since $h = 1$, the series in line 21 to denote the test data is a single element. And that single element corresponds to index $i+T$. Now, when $h > 1$, this part will remain as is since it is possible to make a forecast h periods out. It is the error computations that need to be handled as a matrix that will be explained later for $h > 1$.

Now let us consider line 24 which is the main model used in exponential smoothing. What the line says is to create a model via the exponential smoothing module by offering it the array called `train`. Then for the model itself, consider an additive trend (as we had discussed earlier), and a multiplicative seasonality. Hence the parts `trend = 'add', seasonal = 'mul'`. Then as the final argument of the exponential smoothing function, we provide the number of seasonal periods. Note that since we are inside the "for loop", lines 19 to 36 will run multiple times.

Line 25 essentially asks the exponential smoothing model to train the data and obtain a fit. So if we provide a forecast horizon, it would make a prediction. That is done in line 28. In line 28, we ask the function `fit_model` to make a forecast h periods out. Note that this can also take $h > 1$ and then the `preds` will be an array. However, here for $h = 1$, `preds` is just a single element corresponding to the prediction made for index $i+T$. So `preds` is made using the trained model and making a forecast in line 28.

When $h = 1$, we wish to compare the single predicted value `preds` against the ground truth which is what was observed in the CSV file, i.e., `test`. Now, when $h > 1$, then both `preds` and `test` are h dimensional arrays and we need to compare element by element. Now, if we wish to compute the average error across these h predictions, then we can simply use lines 31 and 32 to get the absolute percentage error and the absolute error respectively. The latter is the absolute difference between the forecast and the ground truth. If $h > 1$ this would be a vector absolute error of size h. But when $h = 1$, it is simply the scalar absolute difference between `preds` and `test`.

Likewise the former is the absolute error over the ground truth, hence percentage error. Of course, if $h > 1$ it is the percentage error of each element in the prediction array, noting that in python when such a division is done, it is term by term of the array. However, when $h > 1$ we may be interested in obtaining the error for each of the h periods out (as is done above). Or we may be interested in just the last one (in situations where we want to make a prediction at a specific time, say 1.25 hours from now). In those case we should not use lines 31 and 32 but instead just take the last element of the array. Then, in lines 35 and 36, we take percentage and absolute errors and append them to the empty list that was created in lines 14 and 15. When $h = 1$, this appended list is just an array of the observed errors. Whereas, when $h > 1$, we need to convert the appended lists into matrices where each row corresponds to a run of the "for loop" and each column, the various time steps in the horizon. This conversion to matrix is in lines 38 and 39.

If we do end up using exponential smoothing, then in practice we would just use the `preds` as our prediction for the next h intervals. But here we are interested in comparing methods and evaluating their performance. So, refer to cells 5 and 6 in Figure 3.7. In cell 5 we use the percentage error list (that we created earlier by appending) and obtain various metrics. We begin with the MAPE (mean absolute percentage error) which in the output is over 13% for this data. However, the median computed in line 3 (and printed next) is 5.7% which is a lot lower than the mean indicating that there are some very large errors but a majority have percentage error lower than 0.057 (as it is the median). The 75th and the 90th percentile are also provided.

In cell 6, as an alternate metric we consider the absolute error ratio. The absolute error by itself does not say much without a baseline or benchmark. We use the average irradiance during the time period in October 2016 that the data was collected as the benchmark (this is computed in line 2 of cell 6). We divide the absolute error array by the average irradiance (which is `input_data['global'].mean()`). Then, similar to cell 5, we take the mean, median, 75th percentile, and 90th percentile. Notice that the mean error ratio is 9% which means that compared to the average irradiance, the average of the absolute value of the error is about 9% (likewise the median is about 5%, the 75th percentile 12%, and 90th percentile is 22%). For $h = 1$, the outputs of cells 5 and 6 are scalars, in other words, arrays with one element.

It is also important to understand how the errors behave based on how far into the future we predict. For this, we simply replace line 8 with $h = 10$ in Figure 3.6. The rest of the code is identical. The output in cells 5 and 6 will now be vectors

```
In [4]:
1  # Obtain solar irradiance data as a time series
2  time_series = input_data['global']
3
4  # Define the parameters for exponential smoothing
5  seasonal_periods = 32
6
7  # Define the number of predictions to make
8  h = 1
9
10 # Define the length of each training set
11 T = 32*7
12
13 # Initialize the lists to store the percentage and absolute errors
14 perc_error_list = []
15 abs_error_list = []
16
17 # Loop through the data frame and make predictions using exponential smoothing
18 for i in range(len(time_series) - T - h + 1):
19     # Define the training and testing data sets
20     train = time_series.iloc[i:i+T]
21     test = time_series.iloc[i+T:i+T+h]
22
23     # Fit the exponential smoothing model
24     model = ExponentialSmoothing(train, trend='add', seasonal='mul', seasonal_periods=seasonal_periods)
25     fit_model = model.fit()
26
27     # Make predictions
28     preds = fit_model.forecast(h)
29
30     # Calculate percentage and absolute errors
31     perc_errors = np.abs(test-preds)/test
32     abs_errors = np.abs(test-preds)
33
34     # Store the percentage and absolute errors
35     perc_error_list.append(perc_errors)
36     abs_error_list.append(abs_errors)
37
38 perc_error_list = np.array(perc_error_list).reshape(len(time_series) - T - h + 1,h)
39 abs_error_list = np.array(abs_error_list).reshape(len(time_series) - T - h + 1,h)
```

Figure 3.6 Cell 4 of the Python Code for Exponential Smoothing

```
In [5]:  # Print the percentage-error results
         print('Mean absolute percentage error:', np.mean(perc_error_list, axis = 0))
         print('Median absolute percentage error:', np.median(perc_error_list, axis = 0))
         print('75th percentile of absolute percentage error:', np.percentile(perc_error_list, 75, axis = 0))
         print('90th percentile of absolute percentage error:', np.percentile(perc_error_list, 90, axis = 0))

         Mean absolute percentage error: [0.13193169]
         Median absolute percentage error: [0.05674817]
         75th percentile of absolute percentage error: [0.13922387]
         90th percentile of absolute percentage error: [0.2924831]

In [6]:  # Print the absolute error ratio results
         avg_global = input_data['global'].mean()
         print('Mean absolute error ratio:', np.mean(abs_error_list, axis = 0)/avg_global)
         print('Median absolute error ratio:', np.median(abs_error_list, axis = 0)/avg_global)
         print('75th percentile absolute error ratio:', np.percentile(abs_error_list, 75, axis = 0)/avg_global)
         print('90th percentile absolute error ratio:', np.percentile(abs_error_list, 90, axis = 0)/avg_global)

         Mean absolute error ratio: [0.0905661]
         Median absolute error ratio: [0.05155128]
         75th percentile absolute error ratio: [0.11583304]
         90th percentile absolute error ratio: [0.21944983]
```

Figure 3.7 Cells 5 and 6 of the Python Code for Exponential Smoothing

Table 3.2

Error metrics based on absolute percentage error (APE) and absolute error ratio (AER) for a python code run with $h = 10$

Metrics	$h=1$	$h=2$	$h=3$	$h=4$	$h=5$	$h=6$	$h=7$	$h=8$	$h=9$	$h=10$
Mean APE	0.132	0.180	0.206	0.228	0.242	0.250	0.264	0.271	0.286	0.304
Median APE	0.057	0.073	0.079	0.096	0.100	0.101	0.108	0.114	0.123	0.131
75th perc. APE	0.140	0.184	0.214	0.242	0.240	0.272	0.288	0.296	0.333	0.361
90th perc. APE	0.292	0.436	0.511	0.553	0.554	0.559	0.605	0.661	0.688	0.725
Mean AER	0.091	0.120	0.136	0.149	0.157	0.165	0.173	0.179	0.190	0.199
Median AER	0.052	0.070	0.076	0.091	0.097	0.095	0.106	0.110	0.115	0.122
75th perc. AER	0.117	0.151	0.178	0.208	0.211	0.227	0.241	0.254	0.264	0.278
90th perc. AER	0.220	0.312	0.356	0.377	0.402	0.403	0.442	0.453	0.470	0.509

of size 10. This is depicted in Table 3.2. Notice that as we move from predicting from 1 to 10 periods out, the quality of our forecasts get weaker. This is across the board on all metrics. However, this is expected as we have a better way of estimating what will happen in the next 15 minutes as opposed to 2 hours from now. The results from Figure 3.7 ought to match those of the first column $h = 1$ in Table 3.2. Upon looking closely there is a subtle difference in the third decimal in some places. This is because when $h = 10$, the test data stop 10 time periods from the end so all 10 predictions can be made (but $h = 1$ in Figure 3.7 uses all but 1 period).

Next we move on to interval forecasts. Although in practice it is common to make point forecasts and use them, it is recommended to accompany a point forecast with a prediction interval. For example, instead of forecasting the solar irradiance an hour from now as 800 W/m^2 as a point forecast, we can also provide an interval such as $[700, 900]$ where 700 is the lower bound and 900 is the upper bound for the irradiance that we may observe. If the prediction interval is based on a 95% probability, it means that there is a 95% chance that the solar irradiance will be between 700 and 900, with an expected value of 800.

Thus, other than the point forecast $\hat{y}_{T+h|T}$, we also provide a 95% prediction interval as $[LB_h, UB_h]$ denoting the interval with a lower bound (LB) and an upper bound (UB) within which there is a 95% chance of falling. Further, although we only considered $h = 1$ in the examples so far, now we consider the case $h > 1$. So, when we obtain point and interval forecast for a horizon h, we obtain it for $1, 2, \ldots, h$ observations ahead. Now, since there is a seasonality and a trend, these intervals would also need to adjust according to those changes. We will keep this analysis to a single T and *not* consider moving window.

The key idea is that the residuals, i.e., $y_{T+h|Y} - \hat{y}_{T+h|T}$ are assumed to be Normally distributed with mean 0 and standard deviation σ_h. Hence, we can say that $y_{T+h|Y}$ will be centered at $\hat{y}_{T+h|T}$ with a spread that is normally distributed with mean 0 and standard deviation σ_h. However, estimating σ_h in closed form is intractable when the seasonality is multiplicative (as in our case). So a simulation approach is used. In fact, the only option at the time of writing this book in `statsmodels` is to use simulation to obtain an interval forecast.

A rough idea of what happens in the back end of `statsmodels` exponential smoothing simulations is as follows. First the model is fit using the training data. Then we need to prescribe how many sample paths to simulate. A sample path is a set of h values that are generated one by one by sampling from the aforementioned normal distribution for $h = 1$. For that σ_h can be estimated as the standard deviation of the residuals in the training data (technically speaking it is the square root of the estimate of the variance of the historic residuals). Thus at each stage, the next step is simulated.

In other words, we obtain a sample of y_{T+1} as $\hat{y}_{T+1} + \varepsilon$ where ε is a sample from Normal distribution with estimated standard deviation for $h = 1$ as $\hat{\sigma}$ based on historical data. Then Equations (3.1)–(3.4), although they are defined only for t between 0 and T, can be extrapolated using the simulated y_{T+1} to obtain $\hat{y}_{T+2}$. Then we take another sample of ε and simulate y_{T+2}, and go through the above steps to get $\hat{y}_{T+3}$. We repeat this process until we get one sample path of y_{T+1} to y_{T+h}.

We repeat the steps in the previous paragraphs for multiple replications. In each replication we obtain a sample path of how the irradiance would look like for a horizon h. Then, for each i from 1 to h we obtain the 2.5th and 97.5th percentile of y_{T+i} based on the sample paths generated. Those values would correspond to the lower and upper bound of the irradiance predictions for i steps into the future. Readers are encouraged to refer to online resources or articles such as Hyndman et al [21] for details on the simulations. Now we explain how this is executed using a Python code.

The first three cells are identical to what we considered earlier. Cell 4 is depicted in Figure 3.8. It would be worthwhile comparing against the previous cell 4 in Figure 3.6. Until line 12, there are no differences except in line 8 which in Figure 3.8 we now consider predicting 16 steps ahead (instead of $h = 1$) and that corresponds to 15-minute intervals into the next 4 hours. For the definitions of the other metrics

```
In [4]:
1  # Obtain solar irradiance data as a time series
2  time_series = input_data['global']
3
4  # Define the parameters for exponential smoothing
5  seasonal_periods = 32
6
7  # Define the number of predictions to make
8  h = 16
9
10 # Define the length of each training set
11 T = 32*7
12
13 # Number of simulations for obtaining prediction intervals
14 num_simulations = 1000
15
16 # Shift in data
17 i = 15
18
19 # Define the training data set
20 train = time_series.iloc[i:i+T]
21
22 # Fit the exponential smoothing model
23 model = ExponentialSmoothing(train, trend='add', seasonal='mul', seasonal_periods=seasonal_periods)
24 fit_model = model.fit()
25
26 # Make predictions and get prediction intervals
27 forecasts = fit_model.forecast(h)
28 simulation_results = fit_model.simulate(h, repetitions=num_simulations)
29 lower_intervals = np.maximum(np.percentile(simulation_results, 2.5, axis = 1),np.zeros(h))
30 upper_intervals = np.percentile(simulation_results, 97.5, axis = 1)
```

Figure 3.8 Cell 4 to Obtain Interval Forecasts for $h = 16$ using Exponential Smoothing

please see earlier description along with Figure 3.6. Next we describe the remaining lines.

Line 14 in the python code cell 4 shown in Figure 3.8 says that we will perform 1000 simulations. This is to get enough sample paths to obtain the 95% interval for the forecasts. Line 17 states that we will consider a period of length T starting with the 15th row, and this is arbitrary (it is more critical in a moving window where i goes from 0 until the row corresponding to the last week of the data set). Line 20 is same as before and at the exact same line (which is a coincidence). Likewise lines 23 and 24 have been explained earlier as those are identical to lines 24 and 25 in Figure 3.6.

Line 27 in the python code cell 4 shown in Figure 3.8 essentially is the set of point forecasts for the entire horizon h. Here we use $h = 16$, so that means we would get forecasts for 1, 2, …, 16 15-minute periods from the last observation in the training data. Line 28 states that we will obtain 1000 sample paths (the `repetitions` denotes that, and in line 14 we said it was 1000). The matrix `simulation_results` has 1000 rows and h columns corresponding to the sample paths and time steps to predict. Lines 29 and 30 compute the 2.5th and 97.5th percentiles (so that what is in between has a probability of 95%) using the 1000 simulated samples.

To see the results of the forecasts and the intervals in Figure 3.8, refer to Figure 3.9. There we first define a dataframe `test_df` which uses the ground truth from the original dataframe `input_data`. Notice that we obtain the observed `global` as well as the time stamp (called datetime) corresponding to the h time steps after the training data ends (shown in Line 2 of 3.9). Then, lines 3 to 5 adds three columns to the dataframe as the point forecast, upper bound of prediction interval, and lower bound of prediction interval. In the output line (see in the bottom part of the figure: `Out[5]:`) see how they are displayed in columns.

```
In [5]:
# Creating a test dataframe with the actuals, forecast, intervals
test_df = input_data.iloc[i+T:i+T+h]
test_df['ES_forecast'] = np.array(forecasts).reshape(h)
test_df['UB_forecast'] = np.array(upper_intervals).reshape(h)
test_df['LB_forecast'] = np.array(lower_intervals).reshape(h)
test_df
```

Out[5]:

	datetime	global	ES_forecast	UB_forecast	LB_forecast
239	2016-10-08 11:45:00	776.16	764.81	938.47	595.97
240	2016-10-08 12:00:00	783.22	773.43	984.79	563.43
241	2016-10-08 12:15:00	763.66	728.88	961.14	509.32

Figure 3.9 Cell 5 to Display Interval Forecasts for $h = 16$ using Exponential Smoothing

```
In [6]:
1 # define function to import viz libraries
2 import plotly
3 plotly.offline.init_notebook_mode(connected=True)
4 from plotly.graph_objs import *
5 from plotly import tools
6 import plotly.graph_objects as go
7 import seaborn as sns
8 from matplotlib import pyplot
9 from pandas.plotting import autocorrelation_plot
```

```
In [7]:
1 # Plot training data with forecast and intervals
2 plot_data = []
3 cols = ['ES_forecast', 'UB_forecast', 'LB_forecast']
4 plot_data.append(go.Scatter(x= input_data['datetime'][i+160:T+i], y= input_data['global'][i+160:T+i],
5                             mode = 'markers', name = 'Training Series'))
6 for y_col in cols:
7     plot_data.append(go.Scatter(x= test_df['datetime'], y= test_df[y_col],
8                      name = '{}'.format(y_col) ))
9 layout = go.Layout(xaxis = dict(title='datetime'), yaxis = dict(title= 'Global Irradiance'),
10                   title = 'Exponential Smoothing Forecasts with Intervals')
11 fig = go.Figure(data= plot_data, layout=layout)
12 fig.update_xaxes(rangebreaks=[dict(bounds=[16, 8], pattern="hour")])
13 plotly.offline.iplot(fig)
```

Figure 3.10 Cells 6 and 7 to Plot Point and Interval Forecasts for $h = 16$ using Exponential Smoothing

Cells 6 and 7 of the Python code shown in Figure 3.10 are used to visualize the training data and the ensuing forecasts (both the point forecasts as well as the prediction interval). First we describe the code and then discuss the output. There are many visualization libraries used in Python. Two popular ones are matplotlib and plotly. Although we will use both in this book, for this exercise we mainly focus on plotly as it allows for significant customization and sophistication compared to matplotlib. That said, matplotlib's pyplot is also imported in line 8 of cell 6 in Figure 3.10. We do not use lines 7 and 9 in this code and will explain later, if needed. Lines 2 to 6 essentially import the necessary plotly functionalities and libraries.

In cell 7 of Figure 3.10, we get into plotting the training time series and also show how the forecasts (point and interval) look like. Let us start by explaining line 4 of cell 7 in Figure 3.10. It essentially asks to create a scatter plot of the irradiance observed in the training data against time. However, we restrict the graph to only

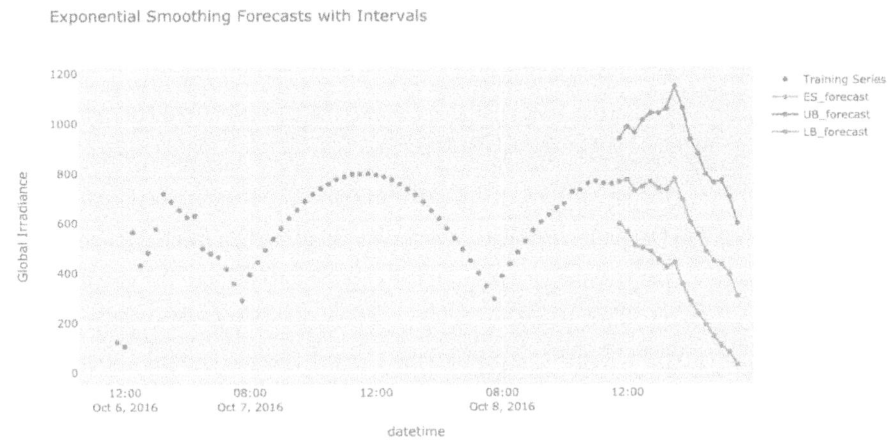

Figure 3.11 Output of Cell 7: Train Data with $h = 16$ point and interval forecasts

2 days (otherwise with all 7 days of training, it would be difficult to appreciate the predictions). So the `i+160` essentially means we start after 32 (number of samples in each day) times 5 (days).

Lines 6–8 do a similar scatter plot except for the three columns (described in line 3) corresponding to the point forecast, upper bound of interval, and lower bound of interval. The idea is to store all these as `plot_data` and plot them. In line 12 we have asked that the hours between 1600 (included) and 800 (not included) be removed. Otherwise those values when there are no data points would simply produce a blank. The rest of the lines are standard verbiage in the plotly library to get the plots, display them, give names to the graph, the axes, and the special lines.

Upon running cell 7, we obtain the output in Figure 3.11. Notice from Figure 3.11 that October 6, 2016 in the location was a day with scattered clouds, and so the irradiance gets badly affected. However, October 7, 2016 seems like a sunny day with a clear sky. Note that we have chopped off the night times. Notice that October 8, 2016 starts clear and gets some minor clouds until 11:30am when the last training data is obtained. Starting 11:45am the forecasts come in for 16 time intervals. Hence the last one is 3:30pm. The top jagged line is the upper bound, the middle jagged line is the point forecast, and the lower jagged line is the lower bound of the interval prediction.

It is important to make a few more observations. As expected, the lines do have a tendency to go down to zero, and that is due to the seasonality observations. Also, since we use data from 7 days to build a model, the effect of October 6, 2016 having a cloudy day can be felt in the forecast. Before going further, it is worthwhile iterating that the 12:00 in the figure is noon, while 08:00 is 8am of the next day (this is the first data point of the day, and the last one is 3:45pm). Upon looking at the values

```
In [8]:
1  # Plot test data to compare forecast versus actuals
2  cols = ['global', 'ES_forecast', 'UB_forecast', 'LB_forecast']
3  plot_data = []
4
5  for y_col in cols:
6      plot_data.append(go.Scatter(x= test_df['datetime'], y= test_df[y_col],
7                                  name = '{}'.format(y_col) ))
8
9  layout = go.Layout(xaxis = dict(title='datetime'), yaxis = dict(title= 'Global Irradiance'),
10                    title = 'Exponential Smoothing Forecasts with Intervals')
11 fig = go.Figure(data= plot_data, layout=layout)
12 plotly.offline.iplot(fig)
```

Figure 3.12 Cell 8 to Obtain Interval Forecasts for $h = 16$ using Exponential Smoothing

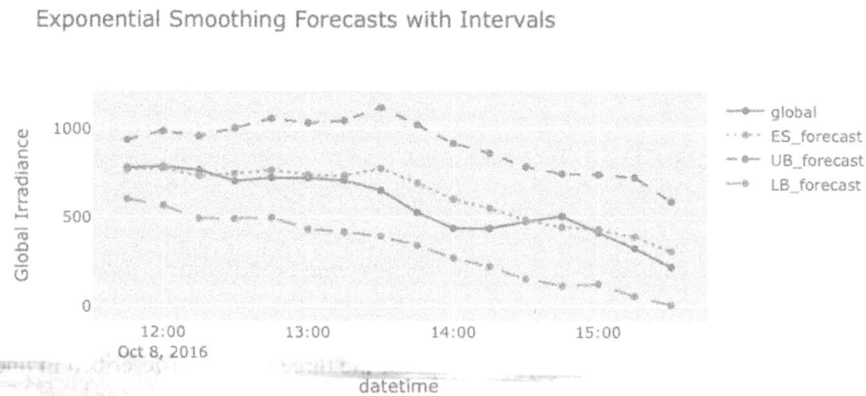

Figure 3.13 Output of Cell 8: Test Data with $h = 16$ point and interval forecasts

forecasted for October 8, 2016, although the line is jagged, there is a high point around 1:30pm. Note that it corresponds to the high point observed on the left on October 6th at 1:30pm. So, exponential smoothing carries the effect of recent days, hence it does look like a combination of the last 2 days.

A natural question to ask is how the forecasts compare with the actuals. For that, we use cell 8 of the python code shown in Figure 3.12. This is a plot of just the test data where we compare the `'global'` which is the ground truth, i.e., the observed irradiance value. We compare that against the point forecast and the interval predictions. Other than that described in line 2 of cell 8, the rest is extremely similar to cell 7. The resulting graph is shown in Figure 3.13. Although not exact, the point forecast and the ground truth are closely tied. Also the upper and lower bounds of the intervals are maintained in this case throughout.

Having seen how to make point and interval forecasts for solar irradiance using exponential smoothing trained on historical time series observation, next we move onto another time-series approach and follow a similar set of steps.

3.2 ARIMA WITH SEASONALITY AND EXOGENOUS VARIABLES

Similar to exponential smoothing, ARIMA is a well-studied time-series forecasting approach. Here too we will use the same notation as in Hyndman and Athanasopoulos [20], like we did with exponential smoothing. The AR in ARIMA is auto-regressive and the MA is moving-average. Hyndman and Athanasopoulos [20] explains them one by one. Here we will directly get into the integrated model (the I in ARIMA is "integrated"). In fact, we will add a seasonal component as part of our explanation. Subsequently, we will bring in exogenous variables. Adding the Seasonal component and the eXogenous variables result in the name SARIMAX. Our goal is to briefly go over SARIMAX and also adapt it for short-term solar forecasting here (and in day-ahead prediction in the next chapter).

We will first explain the ARIMA model with seasonality and show how it can be adapted to solar forecasting. We avoid repeating what was said, so the reader is encouraged to go through Section 3.1 on exponential smoothing to graphically understand how cloud cover (recency), time of day (seasonality), and time of year (trend) affect solar forecasts. The same principle is applied here as well. However, the approach taken in ARIMA is to first remove the effects of seasonality and trend, and then simply use recency to make the forecasts, and then re-introduce the seasonality and trend. Hence, as an initial step, we try to create a *stationary* time-series.

A stationary time-series in an ideal condition is independent of the time stamp. Both trend and seasonality would render a time-series to be non-stationary and they would have to be removed from the data to use the ARIMA method. That is because as said earlier, ARIMA assumes the time-series to be stationary. The approach used to create a "nearly" stationary time-series is called differencing. To explain differencing, recall that we have observations y_0, y_1, ..., y_T that we use for making the predictions into the future. In case of solar forecasts, note that the observations we use are only from 8 hours of a day time-averaged over 15 minutes. So we have 32 samples each day.

For some t in $\{0,1,\ldots,T\}$, let y_t be the tth observation of the time-series. We denote the time series sometimes as $\{y_t : 0 \leq t \leq T\}$. Now consider the time-series $\{y'_t : 1 \leq t \leq T\}$ which is called the first-order differencing term and defined as

$$y'_t = y_t - y_{t-1} \tag{3.6}$$

for all $t \in \{1,2,\ldots,T\}$. Note that the differenced series has one fewer term than the original time-series. Sometimes the series $\{y'_t : 1 \leq t \leq T\}$ may not be adequately stationary, and it may be required to do one more round of differencing.

To perform a second-order differencing, consider the time series $\{y''_t : 2 \leq t \leq T\}$ which is defined as

$$y''_t = y'_t - y'_{t-1} \tag{3.7}$$

for all $t \in \{2,3,\ldots,T\}$. Note that the second-order differenced series has one fewer term than the first-order differenced time-series and two fewer than the original time series. For solar irradiance time series, Figure 3.14 shows three panels corresponding

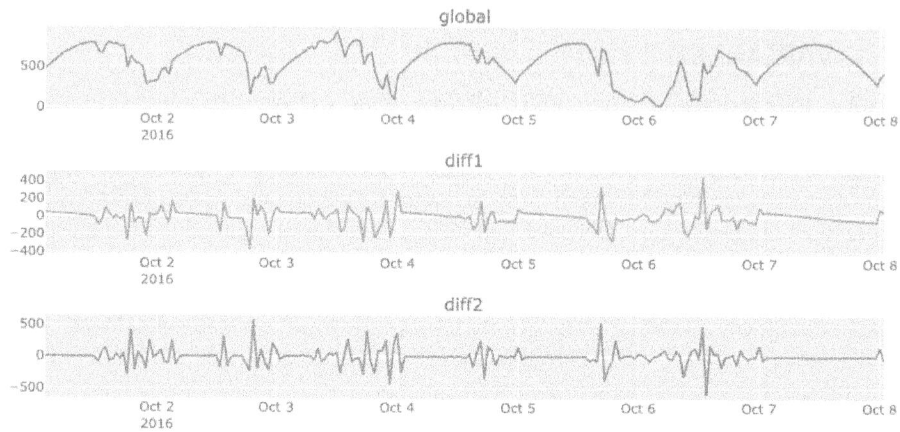

Figure 3.14 Original series, first-order differencing, second order differencing

to the original time series (global), first-order differencing (diff1), and second-order differencing (diff2).

Recall that to plot Figure 3.14, we have explicitly instructed in the python code to remove the times between 4pm and 8am. For this, refer to line 12 in Figure 3.10 under Exponential Smoothing. We do not present the code block in python here as it is similar to that. Now, in Figure 3.14 notice that the first- and second-order differencing, both result in series that look stationary. We will see later that an automated code in fact picks first order. It is generally not recommended to go beyond second-order differencing, although the series may not be stationary. Sometimes that is because of lingering seasonal effects.

When there are seasonal effects, it may be worthwhile (although not guaranteed) to perform seasonal differencing. Let $\{y_t^s : m \leq t \leq T\}$ be lag-m differences, where m is the period of seasonality. The seasonal differencing term is defined as

$$y_t^s = y_t - y_{t-m} \tag{3.8}$$

for all $t \in \{m, m+1, \ldots, T\}$. Then, if needed, we could perform a second-differences (one against season and one against consecutive terms) to obtain the time series $\{y_t^{s\prime} : m+1 \leq t \leq T\}$ which is defined as

$$y_t^{s\prime} = y_t^s - y_{t-1}^s \tag{3.9}$$

for all $t \in \{m+1, m+2, \ldots, T\}$. Note that the notation used in Hyndman and Athanasopoulos [20] for seasonal differencing is different from what is presented above.

In other words, the second-difference in Equation (3.9) will produce the same result whether we take the first-order difference and then the seasonal difference (not shown), or the seasonal difference and then the second differences (as described above). In summary, when we have a time series such as the `global` irradiance data, we need to see if we need to perform differencing and select the best differencing levels. There are statistical tests that can be used. However, instead of doing those tests manually, there are functions in Python that will select the optimal differencing levels. We explain that next.

A commonly used statistical package for ARIMA is `pmdarima`. As a first step it would be necessary to install (such as using the command `pip install pmdarima`). Then use the first three cells that were considered in Exponential Smoothing. Recall in Cell 3 (of Figure 3.5) we created a data frame `input_data`. We use that same data frame in the next cell. Type the following lines of code in the fourth cell:

```
1   import pmdarima as pm
2   D_opt = pm.arima.nsdiffs(input_data['global'], m = 32)
3   d_opt = pm.arima.ndiffs(input_data['global'])
```

where the first line essentially imports the pmdarima library into Python, the second line obtains the optimal seasonal differences `D_opt`, and the third line obtains the optimal regular differences `d_opt`. We can simply type or even use the print command to get the optimal differences. In our case for this dataset, we got 0 and the optimal seasonal differences and 1 for the regular differences. This could change for other datasets and other times of the year.

Now, having obtained an approximately stationary time-series, we explain the ARIMA approach. Again, the reader would find details in Hyndman and Athanasopoulos [20] as we are simply recapitulating the results therein. We begin with ARIMA with no seasonal effects. This allows us to explain without getting bogged down by complex notation. Say we have a differenced series y'_t (this could be the original series y_t if `d_opt` is 0, or the first-order differencing in Equation (3.6) if `d_opt` is 1, or the second-order differencing in Equation (3.7) if `d_opt` is 2).

Under the ARIMA model, the differenced series y'_t is assumed to satisfy

$$y'_t = c + \phi_1 y'_{t-1} + \ldots + \phi_p y'_{t-p} + \theta_1 \varepsilon_{t-1} + \ldots + \theta_q \varepsilon_{t-q} + \varepsilon_t \tag{3.10}$$

for some whole numbers p and q, where $\phi_1, \ldots, \phi_p$ are the p coefficients of the AR terms, $\theta_1, \ldots, \theta_q$, the q coefficients of the MA terms, c the equivalent of the intercept in a regression model, and ε_t is white noise (that effectively corresponds to the residuals in which case it is a function of the coefficients as well as the time series values) for all t.

The above is known as an **ARIMA**(p,d,q) model as it has p as the number of autoregressive terms, d the degree of differencing (such as the result `d_opt`) and q the number of moving average terms. In Equation (3.10), we do not know the parameters c, $\phi_1, \ldots, \phi_p$, $\theta_1, \ldots, \theta_q$, and they have to be estimated. Also, the ε_t terms are a function of c, $\phi_1, \ldots, \phi_p$ as well as the time series. To illustrate this, let us take the case of $p = 1$, $d = 0$, and $q = 1$. Then the model becomes

$$y_t = c + \phi_1 y_{t-1} + \theta_1 \varepsilon_{t-1} + \varepsilon_t$$

for $t = 1, \ldots, T$. The parameters c, ϕ_1, and θ_1, need to be estimated.

There are two approaches to obtain the parameters, one is to minimize the sum of squares of ε_t, and the other is to maximize the likelihood of obtaining the series y'_t for all t assuming it is Gaussian and so are the ε_t values. Either case would result in a complicated non-linear optimization problem that cannot be written in closed-form. The details of the methods are beyond the scope of this book. However, there are numerous software packages that produce the results. In the ARIMA model described in Equation (3.10), we typically use Maximum Likelihood Estimator (MLE) to estimate c, $\phi_1, \ldots, \phi_p, \theta_1, \ldots, \theta_q$. The default setting is MLE in the Python libraries we will show here.

Before going to the Python code, it is worthwhile to discuss the seasonal ARIMA model and to explain the code for the seasonal case. The key idea of seasonal-ARIMA is to add seasonal terms and extend the **ARIMA**(p,d,q) model to the **seasonal-ARIMA**$(p,d,q)(P,D,Q)$ with m seasons. Refer to Section 3.1 for a detailed explanation of how $m = 32$ as the number of seasonal periods for the 15-minute average solar irradiance data with 8 hour daylight windows. Although we do not present the equation, it is worthwhile understanding that to Equation (3.10) we add seasonality AR terms $y_{t-m}, \ldots. y_{t-Pm}$ with coefficients $\Phi_1, \ldots, \Phi_P$, as well as seasonality MA terms $\varepsilon_{t-m}, \ldots, \varepsilon_{t-Qm}$ with coefficients $\Theta_1, \ldots, \Theta_Q$.

Although we have discussed (but not shown) how to get the coefficients $\phi_1, \ldots, \phi_p$, $\theta_1, \ldots, \theta_q, \Phi_1, \ldots, \Phi_P, \Theta_1, \ldots, \Theta_Q$ using MLE, we also need to determine how to get p, q, P, and Q. They can be selected by searching through the space of values and find the one that minimizes an information criterion. The common choice of information criteria are AIC (Akaike's Information Criterion), AICc (i.e., AIC corrected), and BIC (Bayesian Information Criterion). We do not present the expressions here but they are typically functions of likelihood of obtaining the data (that was used in MLE), T, p, q, P, Q, and whether or not c was used.

Although we would use the following `pmdarima` based code block later while running an ARIMA model (with seasonality) code, it is worthwhile showing some of the various features that are commonly used. Consider the following cell:

```
1   import pmdarima as pm
2   from pmdarima import auto_arima
3   model = auto_arima(y=train, start_p = 1, d = 1, start_q = 2,
4                      max_p = 3, max_q = 4, start_P = 0, start_Q = 0,
5                      max_P = 0, max_Q = 1, max_D = 1, m = 32)
```

which would be right after a time series called `train` is obtained. Lines 1 and 2 in the above cell essentially imports `pmdarima` and also `auto_arima` built in `pmdarima`. Lines 3–5 state that we would fit a model using `auto_arima`, and what that means is the function would first select the best options for p, d, q, P, D, and Q using AIC and then obtain the coefficients $\phi_1, \ldots, \phi_p, \theta_1, \ldots, \theta_q, \Phi_1, \ldots, \Phi_P, \Theta_1, \ldots, \Theta_Q$.

The above is just an example, and depending on the problem they as well as others that have been set to their defaults can be selected. Then the function `auto_arima` starts with an initial choice of p, q, P, and Q called `start_p`, `start_q`, `start_P`, and `start_Q`, and then searches through the space of values which have an upper bound `max_p`, `max_q`, `max_P`, and `max_Q`. When d is not specified, it searches for one (and that could be time-consuming). In the code block above, we set it to 1 (in this case that is chosen using `ndiffs` described earlier to select `d_opt`). Likewise D as well can be pre-specified. In the above lines we did not do that but restricted it to a maximum of 1. Finally we also specify the seasonal periods m.

Once the model is built, we use the fitted parameters to predict or forecast for the next several time intervals (in case of multiple intervals, we just assume that the forecasted value is the actual value). Recall that we have observations $y_0, y_1, \ldots, y_T$. We wish to estimate y_{T+h} which is written as $\hat{y}_{T+h|T}$. In other words, we are making an h-step ahead forecast. For the first set of results we will use $h = 1$ meaning the very next time period. After that we will consider $h > 1$. Also, we tend to use a moving window. So first we predict y_{T+h} using 0 to T, then y_{T+1+h} with 1 to $T+1$, followed by predicting y_{T+2+h} using 2 to $T+2$, and so on. Although it is not necessary, we assume that we will train the model for each of the time windows.

That said, we now present the code for ARIMA with seasonality. We use the same words as in ES, but when needed adapted to ARIMA. We use the same T as in ES and the same datasets so that we could compare the methods. As stated in ES, a frequently used rule of thumb is about 200 points to train as T. For 15-minute granularity data taken for 10 hours in a day, the training is about 5 days. However, the training period T would be about 20 days for 1-hour granularity. In our code, we first only provide point forecasts. That means we provide the expected value of what the forecast would be. After going through the point forecast code, we will discuss obtaining a forecast interval.

To obtain a point forecast of solar irradiance, we use a Python code and provide the cells of a Jupyter notebook. The first cell is identical to that in Figure 1.6. In Figure 3.15 we provide the second cell (represented as `In [2]:` in Figure 3.15). There are three lines in that cell marked from 1 to 3. Line 1 is simply a comment that says what the cell does. In line 2 we import the `pmdarima` package and call it `pm`. In line 3, we take `auto_arima` classes and functions from the python module `pmdarima`. Next we move to the third cell (represented as `In [3]:` in Figure 3.15). There we use `pandas` written in short form as pd (see Figure 1.6) to read from a CSV file. The file is read into a dataframe `input_data` in line 2 of the third cell. To display the header of the data frame, we write down line 3.

Notice the output of the third cell (represented as `Out [3]:` in Figure 3.15). The leftmost column without a name but with numbers 0, 1, 2, 3, and 4 is the index of the dataframe. Then come the two named columns of the dataframe: datetime and global. The datetime column gives the date and time (in format YYYY-MM-dd hh:mm:ss, where YYYY denotes the four-digit year, MM denotes month in 2 digits, dd stands for day, then a space followed by the hours, minutes and seconds). Notice that the minutes go in steps of 15, that is because the data is averaged over 15-minute windows. The column called "global" is the solar irradiance in W/m^2 that is averaged over a fifteen-minute interval.

Next we move to the main Python code in the Jupyter notebook cell 4 which (represented as `In [4]:`) is the entire Figure 3.16. The lines that begin with the hashtag symbol are comments that essentially explain what is coming. Hence those

```
In [2]:  # Key imports for this code
         import pmdarima as pm
         from pmdarima import auto_arima
```

```
In [3]:  # Obtain solar irradiance data as a data frame
         input_data = pd.read_csv('Oct2016_no_night_solarTAC.csv', parse_dates = ['datetime'])
         input_data.head()
```

Out[3]:

	datetime	global
0	2016-10-01 08:00:00	392.13
1	2016-10-01 08:15:00	439.06
2	2016-10-01 08:30:00	483.33
3	2016-10-01 08:45:00	525.62
4	2016-10-01 09:00:00	566.48

Figure 3.15 Cells 2 and 3 of the Python Code for ARIMA

lines are not presented and our focus is on what the non-commented lines indicate. In line 2, we create an array called `time_series` by taking the column "global" from the input data frame (called `input_data`). The reason for this is that the time-series needs to be presented as an array (and not a data frame element) when passed on to the ARIMA function.

Line 5 `seasonal_periods` denotes the *period of seasonality* that we denoted as m earlier. This is 32 because each day has 8 hours of data (where there is significant sunshine) and each hour has 4 data-points as the granularity is 15 minute average. Recall (from the CSV file read in Figure 3.15) that the CSV file only has daytime data and low solar power or dark times have been removed. So, every 32 data points denote the same time but a day apart. In line 8 we say that we make an h-step-ahead forecast for $h = 1$. Then in line 11 we use a 7-day history to train a model and make a prediction h days after.

The code itself has been written for $h = 1$, and no further modifications will be needed for $h > 1$. We will explain the implications of $h > 1$ later as needed. Lines 14 and 15 are empty lists that would be populated with the percentage error and the absolute error made for each forecast at $h = 1$ steps ahead. We will populate the lists in lines 31 and 32. Now we focus on line 18 for the case $h = 1$. This is the part where we do a moving window. We first train using 0 to $T - 1$ and predict the irradiance for observation T. Then we move the window by 1 and train using 1 to T to predict $T + 1$. Next we use 2 to $T + 1$ to predict $T + 2$. We keep doing this until we run out of elements in the time series array.

Thus in line 18, we start with index 0 and go until the last possible index which is the length of the time series minus $T + 1$ so that each line in the `for loop` selects a training data of length T. Note that in Python the indexing is from 0 to $T - 1$ (as opposed to 1 to T). Also the function `range()` lists all the numbers from 0 until the number in the argument minus 1. So `range(5)` would be 0, 1, 2, 3, 4. Hence the "for loop" in line 18 essentially says that i starts from 0 and goes all the way until the last possible index that would leave at least 1 data point in the set that we can compare against while making the prediction.

In other words, i is the first index in the set of indices of length T that is used to train the data. This is depicted in line 20. The series `train` in line 20 is the array corresponding to elements indexed from i to $i+T-1$ (this is written as $[i:i+T]$). Since $h=1$, the series in line 21 to denote the test data is a single element. And that single element corresponds to index $i+T$. Now, when $h>1$, this part will remain as is since it is possible to make a forecast h periods out. It is the error computations that need to be handled as a matrix that will be explained later for $h>1$.

Now let us consider line 24 which is the main model used in ARIMA. What the line says is to create a model via `auto_arima`. We explained earlier how the function works. However, here we use a different set of values and only those are explained here. For example, we do not provide any values of p, d, and q but let `auto_arima` select the best. However, the code runs slowly especially in this interactive "for loop" of training and prediction. So, we use $P=0$, $D=0$, and try Q values of 0 and 1. We also provide m equal to the number of seasonal periods. Note that since we are inside the "for loop", lines 19 to 36 will run multiple times.

Line 28 essentially asks the ARIMA model to use the trained data to make a prediction. In line 28, we ask the function `fit_model` to make a forecast h periods out. Note that this can also take $h>1$ and then the `preds` will be an array. However, here for $h=1$, `preds` is just a single element corresponding to the prediction made for index $i+T$. So `preds` is made using the trained model and making a forecast in line 28. Now, also notice that we ask the code to not return a confidence interval as we are only making a point forecast now. We will obtain an interval later.

When $h=1$, we wish to compare the single predicted value `preds` against the ground truth which is what was observed in the CSV file, i.e., `test`. Now, when $h>1$, then both `preds` and `test` are h dimensional arrays and we need to compare element by element. Now, if we wish to compute the average error across these h predictions, then we can simply use lines 31 and 32 to get the absolute percentage error and the absolute error respectively. The latter is the absolute difference between the forecast and the ground truth. If $h>1$ this would be a vector absolute error of size h. But when $h=1$, it is simply the scalar absolute difference between `preds` and `test`.

Likewise the former is the absolute error over the ground truth, hence percentage error. Of course, if $h>1$ it is the percentage error of each element in the prediction array, noting that in python when such a division is done, it is term by term of the array. However, when $h>1$ we may be interested in obtaining the error for each of the h periods out (as is done above). Or we may be interested in just the last one (in situations where we want to make a prediction at a specific time, say 1.25 hours from now). In those cases we should not use lines 31 and 32 but instead just take the last element of the array. Then, in lines 35 and 36, we take percentage and absolute errors and append them to the empty list that was created in lines 14 and 15. When $h=1$, this appended list is just an array of the observed errors. Whereas, when $h>1$, we need to convert the appended lists into matrices where each row corresponds to a run of the "for loop" and each column, the various time steps in the horizon. This conversion to matrix is in lines 38 and 39.

```
In [4]:
# Obtain solar irradiance data as a time series
time_series = input_data['global']

# Define the parameters for ARIMA
seasonal_periods = 32

# Define the number of predictions to make
h = 1

# Define the length of each training set
T = 32*7

# Initialize the lists to store the percentage and absolute errors
perc_error_list = []
abs_error_list = []

# Loop through the data frame and make predictions using auto ARIMA
for i in range(len(time_series) - T - h + 1):
    # Define the training and testing data sets
    train = time_series.iloc[i:i+T]
    test = time_series.iloc[i+T:i+T+h]

    # Fit the ARIMA model
    model = auto_arima(y=train, m = seasonal_periods,
                       start_P = 0, start_Q = 0, max_P = 0, max_Q = 1, D = 0)

    # Make predictions
    preds = model.predict(n_periods=h, return_conf_int=False)

    # Calculate percentage and absolute errors
    perc_errors = np.abs(test-preds)/test
    abs_errors = np.abs(test-preds)

    # Store the percentage and absolute errors
    perc_error_list.append(perc_errors)
    abs_error_list.append(abs_errors)

perc_error_list = np.array(perc_error_list).reshape(len(time_series) - T - h + 1,h)
abs_error_list = np.array(abs_error_list).reshape(len(time_series) - T - h + 1,h)
```

Figure 3.16 Cell 4 of the Python Code for ARIMA

If we do end up using ARIMA, then in practice we would just use the `preds` as our prediction for the next h intervals. But here we are interested in comparing methods and evaluating their performance. So, refer to cells 5 and 6 in Figure 3.17. In cell 5 we use the percentage error list (that we created earlier by appending) and obtain various metrics. We begin with the MAPE (mean absolute percentage error) which in the output is over 15% for this data. However, the median computed in line 3 (and printed next) is 6.2% which is a lot lower than the mean indicating that there are some very large errors but a majority have percentage error lower than 0.062 (as it is the median). The 75th and the 90th percentile are also provided.

In cell 6, as an alternate metric we consider the absolute error ratio. The absolute error by itself does not say much without a baseline or benchmark. We use the average irradiance during the time period in October 2016 that the data was collected as the benchmark (this is computed in line 2 of cell 6). We divide the absolute error array by the average irradiance (which is `input_data['global'].mean()`). Then, similar to cell 5, we take the mean, median, 75th percentile, and 90th percentile. Notice that the mean error ratio is 9.6% which means that compared to the average irradiance, the average of the absolute value of the error is about 9.6% (likewise the

```
In [5]:
# Print the percentage-error results
print('Mean absolute percentage error:', np.mean(perc_error_list, axis = 0))
print('Median absolute percentage error:', np.median(perc_error_list, axis = 0))
print('75th percentile of absolute percentage error:', np.percentile(perc_error_list, 75, axis = 0))
print('90th percentile of absolute percentage error:', np.percentile(perc_error_list, 90, axis = 0))

Mean absolute percentage error: [0.15205984]
Median absolute percentage error: [0.06254146]
75th percentile of absolute percentage error: [0.17346707]
90th percentile of absolute percentage error: [0.33727274]
```

```
In [6]:
# Print the absolute error ratio results
avg_global = input_data['global'].mean()
print('Mean absolute error ratio:', np.mean(abs_error_list, axis = 0)/avg_global)
print('Median absolute error ratio:', np.median(abs_error_list, axis = 0)/avg_global)
print('75th percentile absolute error ratio:', np.percentile(abs_error_list, 75, axis = 0)/avg_global)
print('90th percentile absolute error ratio:', np.percentile(abs_error_list, 90, axis = 0)/avg_global)

Mean absolute error ratio: [0.09643946]
Median absolute error ratio: [0.06204152]
75th percentile absolute error ratio: [0.11393319]
90th percentile absolute error ratio: [0.22636473]
```

Figure 3.17 Cells 5 and 6 of the Python Code for ARIMA

median is about 6%, the 75th percentile is 11%, and 90th percentile is 23%). For $h = 1$, the outputs of cells 5 and 6 are scalars, in other words, arrays with one element. For the most part, these results are comparable to exponential smoothing in Figure 3.7 and slightly worse in a majority of instances.

Next we try and understand how the errors behave based on the forecast horizon. For this, we simply replace line 8 with $h = 10$ in Figure 3.16. The rest of the code is identical (except line 24 which is explained below). The output in cells 5 and 6 will now be vectors of size 10. This is depicted in Table 3.3. Notice that as we move from predicting from 1 to 10 periods out, the quality of our forecasts get weaker. This is across the board on all metrics. However, this is expected as we have a better way of estimating what will happen in the next 15 minutes as opposed to 2 hours from now. The results from Figure 3.17 are supposed to match those of the first column $h = 1$ in Table 3.3. But that does not happen because on line 24, we added `d = 1,` in the hope of speeding up the code by fixing the differencing. Nonetheless, the results are close enough. In addition, when $h = 10$, the test data stop 10 time period from the end so all 10 predictions can be made (but $h = 1$ in Figure 3.17 uses all but 1 period).

Before forging ahead, it is worthwhile noting that in Table 3.3 the errors are high, especially comparing to the corresponding table that uses exponential smoothing (see Table 3.2). This does not mean that exponential smoothing would always result in better results. In fact, in this situation, the issue may be more with `auto_arima` which some have warned to not just use blindly. In fact, one would have to also check the inputs. For example, using $D = 0$ was perhaps not a good idea as the data has a clear seasonal pattern. Also, the default criteria is AIC but we may wish to consider others. Sometimes the default algorithm does not converge. The main lesson is to check and see what parameters are used (this can be done with the command `model.get_params()` assuming the model name is "model"). In the next few models, we will use fixed parameters for p, d, q, P, D, Q.

Next we move on to interval forecasts using the exact same verbiage as in exponential smoothing. Although in practice it is common to make point forecasts and test them, it is recommended to accompany a point forecast with a prediction

Table 3.3
ARIMA error metrics based on absolute percentage error (APE) and absolute error ratio (AER) for a python code run with $h = 10$

Metrics	$h=1$	$h=2$	$h=3$	$h=4$	$h=5$	$h=6$	$h=7$	$h=8$	$h=9$	$h=10$
Mean APE	0.147	0.232	0.286	0.336	0.375	0.412	0.459	0.497	0.533	0.567
Median APE	0.071	0.117	0.163	0.204	0.235	0.270	0.306	0.334	0.351	0.383
75th perc. APE	0.167	0.259	0.350	0.388	0.432	0.496	0.534	0.587	0.632	0.654
90th perc. APE	0.315	0.475	0.581	0.667	0.716	0.770	0.878	0.962	1.100	1.204
Mean AER	0.100	0.154	0.194	0.228	0.258	0.286	0.316	0.342	0.367	0.388
Median AER	0.066	0.116	0.161	0.199	0.230	0.259	0.297	0.323	0.346	0.366
75th perc. AER	0.115	0.183	0.251	0.306	0.361	0.407	0.453	0.500	0.544	0.574
90th perc. AER	0.225	0.348	0.406	0.449	0.504	0.547	0.602	0.645	0.673	0.709

interval. For example, instead of forecasting the solar irradiance an hour from now as 800 W/m^2 as a point forecast, we can also provide an interval such as $[700, 900]$ where 700 is the lower bound and 900 is the upper bound for the irradiance that we may observe. If the prediction interval is based on a 95% probability (which is the default if nothing is specified), it means that there is a 95% chance that the solar irradiance will be between 700 and 900, with an expected value of 800.

Thus, other than the point forecast $\hat{y}_{T+h|T}$, we also provide a 95% prediction interval as $[LB_h, UB_h]$ denoting the interval between a lower bound (LB) and an upper bound (UB) within which there is a 95% chance of falling. We will consider an h so that $h > 1$. When we obtain point and interval forecast for a horizon h, we obtain it for 1, 2, ..., h observations ahead. We will keep this analysis to a single T and *not* consider moving window. The confidence interval is provided by all packages that do ARIMA with seasonality with the assumption that the residuals, i.e., $y_{T+h|Y} - \hat{y}_{T+h|T}$ are Normally distributed with mean 0 and standard deviation σ_h (which is estimated, a description of the methodology is beyond the scope of this book). Hence we can say that $y_{T+h|Y}$ will be centered at $\hat{y}_{T+h|T}$ with a spread that is normally distributed with mean 0 and standard deviation σ_h.

Now, let us focus on the python code. The first three cells are identical to what we considered earlier (see Figures 1.6 and 3.15). The fourth cell is subtly different and we go into some detail here. The new cell 4 is in Figure 3.18. It would be worthwhile comparing against the previous cell 4 in Figure 3.16. Until line 12, there are no differences except in line 8 which in Figure 3.18 we now consider predicting 16 steps ahead (instead of $h = 1$) and that corresponds to 15-minute intervals into the next 4 hours. For the definitions of the other metrics please see earlier description along with Figure 3.16. Next we describe the remaining lines.

Line 14 in the python code cell 4 shown in Figure 3.18 states that we will consider a period of length T starting with the 15th row, and this is arbitrary (it is more critical in a moving window where i goes from 0 until the row corresponding to the last week of the data set). Line 17 is same as before and essentially creates the T

```
In [4]:  1 # Obtain solar irradiance data as a time series
         2 time_series = input_data['global']
         3
         4 # Define the parameters for ARIMA
         5 seasonal_periods = 32
         6
         7 # Define the number of predictions to make
         8 h = 16
         9
        10 # Define the length of each training set
        11 T = 32*7
        12
        13 # Shift in data
        14 i = 15
        15
        16 # Define the training data set
        17 train = time_series.iloc[i:i+T]
        18
        19 # Selecting an order
        20 order = (3, 1, 0)
        21 seasonal_order = (0, 1, 1, seasonal_periods)
        22
        23 model = pm.arima.ARIMA(order=order, seasonal_order=seasonal_order, method='bfgs')
        24 fit_model = model.fit(y=train)
        25 forecasts, conf_int = fit_model.predict(n_periods=h, return_conf_int=True)
        26 lower_intervals = conf_int[:,0]
        27 upper_intervals = conf_int[:,1]
```

Figure 3.18 Cell 4 to Obtain Interval Forecasts for $h = 16$ using ARIMA with Seasonality

period long training interval. Line 20 states that (p,d,q) will equal $(3,1,0)$ while line 21 says $P = 0$, $D = 1$, $Q = 1$, and the number of seasonal periods defined earlier is the seasonality. The choice of (p,d,q) and (P,D,Q) are somewhat arbitrary, in that they have not been optimized. The idea for using $d = 1$ and $D = 1$ is that we expect some daily seasonality and a second difference could smooth out any small trends.

In line 23, we consider a different model which is also from `pmdarima` like `auto_arima`. We specify (p,d,q) and (P,D,Q) as the first two arguments. Usually the last one is not needed. However, the default optimization method we used to find the parameters that minimize the negative of the logarithm of likelihood did not converge. However, the `bfgs` method did converge. Line 24 states that we use the train data to fit the model and obtain the model parameters. Then in line 25 we ask the model to make predictions h periods out. In that process we wish to obtain both the point estimates (`forecasts`) and the confidence interval (for which we need to specify that the confidence interval must be returned). Then in line 26 and 27 we obtain the lower and upper confidence interval by using the appropriate element in the confidence interval.

To see the results of the forecasts and the intervals in Figure 3.18, we create cell 5 but it is not presented here as it is extremely similar to Figure 3.9 and we simply replace “ES” with “SARIMA”. We do the same switch in cells 6 and 7 shown in Figure 3.10. In cell 5, we define a dataframe `test_df` which uses the ground truth from the original dataframe `input_data`. Notice that we obtain the observed `global` as well as the time stamp (called datetime) corresponding to the h time steps after the training data ends (shown in Line 2 of Figure 3.9). Then, lines 3 to 5 adds three columns

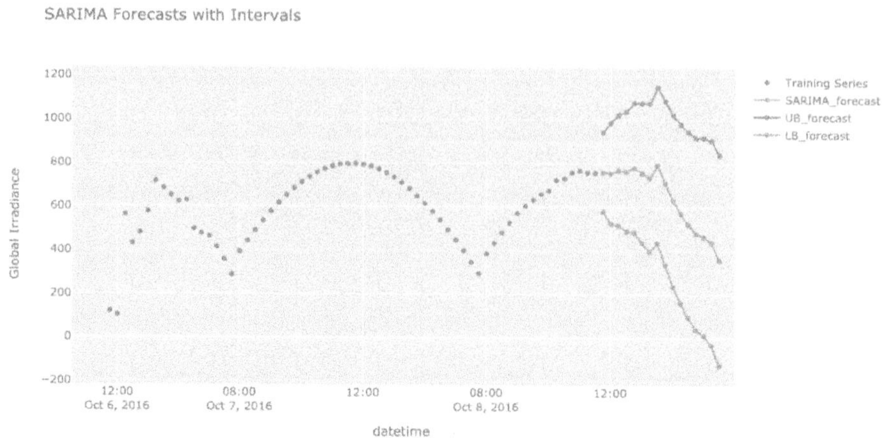

Figure 3.19 Modified Cell 7 Output: Train Data with $h = 16$ point and interval forecasts

to the dataframe as the point forecast, upper bound of prediction interval, and lower bound of prediction interval.

Cells 6 and 7 of the Python code shown in Figure 3.10 (which are modified and used here for SARIMA). The details of the lines of code are along with the description of ES where the figure appears. Upon running cell 7, we obtain the output in Figure 3.19. Notice from Figure 3.19 that October 6, 2016 in the location was a day with scattered clouds, and so the irradiance was badly affected. However, October 7, 2016 seems like a sunny day with a clear sky. Notice that we have chopped off the night times. See that October 8, 2016 starts clear and gets some minor clouds until 11:30am when the last training data is obtained. Starting 11:45am the forecasts come in for 16 time intervals. Hence the last one is 3:30pm. The top jagged line is the upper bound, the middle jagged line is the point forecast and the lower jagged line is the lower bound of the interval prediction.

It is important to make a few more observations. As expected, the lines do have a tendency to go down to zero, and that is due to the seasonality observations. Also, since we use data from 7 days to build a model, the effect of October 6, 2016 having a cloudy day can be felt in the forecast. Before going further, it is worthwhile iterating that the 12:00 in the figure is noon, while 08:00 is 8am of the next day (this is the first data point of the day, and the last one is 3:45pm). Upon looking at the values forecasted for October 8, 2016, although the line is jagged, there is a high point around 1:30pm. Note that it corresponds to the high point observed on the left on October 6th at 1:30pm. So, ARIMA also carries the effect of recent days, hence it does look like a combination of the last 2 days.

A natural question to ask is how the forecasts compare with the actuals. For that, we use cell 8 of the python code shown in Figure 3.12 by making suitable modification from ES or exponential smoothing to ARIMA. This results in a plot of just

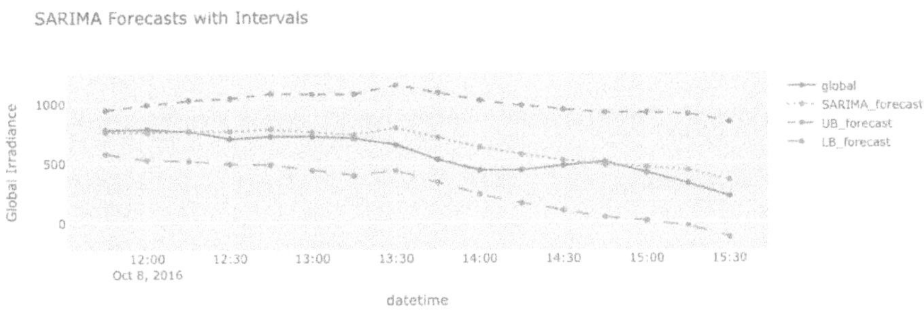

Figure 3.20 Modified Cell 8: Test Data with $h = 16$ point and interval forecasts

the test data where we compare the `'global'` which is the ground truth, i.e., the observed irradiance value. We compare that against the point forecast and the interval predictions. Other than that described in line 2 of cell 8, the rest is extremely similar to the cell 7. The resulting graph is shown in Figure 3.20. Although not exact, the point forecast and the ground truth are closely tied. Also the upper and lower bounds of the intervals are maintained in this case throughout.

One thing to notice is how similar the results are to exponential smoothing. In some sense both models although different in a lot of ways, philosophically do some similar things: they use recent true values, fit trends and seasonality, and then use regressions to get parameters which are then used for predictions. Having seen how to make point and interval forecasts for solar irradiance using SARIMA through a time series of just the irradiance data, next we ask if an how we can use exogenous variables. Exponential smoothing does not allow exogenous variables, but SARIMA does and for that reason it is called SARIMAX (SARIMA with exogenous variables).

We begin by briefly explaining how SARIMAX works. Recall that in ARIMA, we showed in Equation (3.10) how the differenced series y'_t is assumed to be related to the previous observations as well as the residuals. In SARIMAX we assume that y'_t also depends on some exogenous variables such as temperature, wind-speed, and wind direction. To account for the exogenous features, all we do is simply add the exogenous variables each pre-multiplied by corresponding coefficients in Equation (3.10). Then the SARIMAX function (that we will see later) will estimate the coefficients as before using an optimization approach to maximize the likelihood of obtaining the series y'_t for all t assuming it is Gaussian and so are the ε_t values.

With that understanding, we are in a position to explain the SARIMAX code for python. For SARIMAX we use a different data set that also includes some exogenous variables as well as a different python library. These are shown in Figure 3.21 which are cells 2 and 3 of the python code (the first cell as always remains the same). In cell 2, notice that we are considering a different library. Although `pmdarima`'s `auto_arima` does allow for exogenous variables, we take the opportunity to introduce a different library. In lines 2 and 3 of cell-2 we import SARIMAX and the API

In [2]:
```
# Key imports for this code
from statsmodels.tsa.api import ExponentialSmoothing
from sklearn.metrics import mean_absolute_percentage_error
from sklearn.metrics import mean_absolute_error
```

In [3]:
```
# Obtain solar irradiance data as a data frame
input_data = pd.read_csv('Oct2016_no_night_solarTAC.csv', parse_dates = ['datetime'])
input_data.head()
```

Out[3]:

	datetime	global
0	2016-10-01 08:00:00	392.13
1	2016-10-01 08:15:00	439.06
2	2016-10-01 08:30:00	483.33
3	2016-10-01 08:45:00	525.62
4	2016-10-01 09:00:00	566.48

Figure 3.21 Cells 2 and 3 of the Python Code for SARIMAX

for `statsmodels` respectively. Recall that we used `statsmodels` in exponential smoothing. In cell-3 we upload a different dataframe and its header is shown as well.

Let us take a moment to understand that data set before forging ahead. Recall from the beginning of this chapter that the original data is from SolarTAC and obtained from NREL [34] (with data source credited to Andreas and Wilcox [3]). It is from a city called Aurora, Colorado (which is located to the east of the city of Denver). The latitude, longitude, and elevation of SolarTAC are in Table 3.4. As an exogenous variable, we consider the solar irradiance at a location several miles to the West of SolarTAC. This location is SRRL in a city called Golden in Colorado. Golden is to the west of Denver. Solar irradiance data for SRRL is also available from NREL [34] (with data source credited to Andreas and Stoffel [2]). The latitude, longitude, and elevation of SRRL are in Table 3.4. Notice from the table that the elevations are similar and so are the latitudes. However, notice from the longitude that SRRL is more than half a degree west of SolarTAC.

Now refer to Figure 3.21 cell 3 output. The last two columns represent the solar irradiance and the air temperature at SRRL. However, most importantly, they correspond to conditions a time 30 minutes prior to the time in the first column. So that

Table 3.4

Geographic details of the main location (SolarTAC) and the location for exogenous data SRRL

Site	SolarTAC	SRRL
Location	Aurora, CO	Golden, CO
Latitude	39.757° North	39.742° North
Longitude	104.620° West	105.18° West
Elevation	1674m	1829m

```
In [4]:
1 # Obtain solar irradiance data as a time series
2 time_series = input_data['global']
3
4 # Define the parameters for ARIMA
5 seasonal_periods = 32
6
7 # Define the number of predictions to make
8 h = 1
9
10 # Define the length of each training set
11 T = 32*7
12
13 # Selecting an order
14 order = (2, 1, 0)
15 seasonal_order = (0, 1, 1, seasonal_periods)
16
17 # Initialize the lists to store the percentage and absolute errors
18 perc_error_list = []
19 abs_error_list = []
```

Figure 3.22 Cell 4 of the Python Code for SARIMAX

8am data in the table actually corresponds to 7:30am at SRRL, likewise the 8:45am line is data from 8:15am in SRRL. The key *idea* is that it takes about 30 minutes for clouds to move from SRRL to SolarTAC on average. So, to predict irradiance due to a quick cloud cover that would move in half an hour, it would be useful to obtain the conditions half hour earlier. In that region the predominant wind direction is west to east and the average wind speed is such that it takes half hour for the clouds to move. This would need to be done carefully for other locations.

In fact, ideally features such as wind-speed and wind-direction can also be used for the prediction, but that analysis does get complicated. Further, notice we also added air temperature in degrees Kelvin. This is both at SolarTAC and SRRL (which like irradiance is 30 minutes prior). The understanding is that temperatures can be predicted extremely accurately and can serve as useful exogenous variables. For other locations it would be advisable to select an appropriate place where solar irradiance can be measured a few minutes in advance based on wind speed and direction. Note that going too far would result in clouds diffusing, and too close would not be useful. However, one can use a sensor to measure the irradiance and not rely on solar panels which are used in the data set we showed.

Having described cells 2 and 3, we now move to cell 4 where we define the numerical parameters. Refer to cell 4 in Figure 3.22. Notice in line 2 we use the main time series like we did for ARIMA earlier. Lines 5, 8, and 11 are exactly the same as the corresponding ones in ARIMA. In line 14 we choose (p,d,q) as the `order`. The choice of (2, 1, 0) is not necessarily optimized but uses the fact that the previous two irradiance values ($p = 2$) are perhaps sufficient and we did well with $d = 1$ earlier. Further, the choice of $P = 0$, $D = 1$, $Q = 1$ in line 15 is totally arbitrary. In practice, this would need to be carefully selected by optimizing for the information criteria, as described earlier. Lines 18 and 19 initialize the error lists as we did earlier for ARIMA.

```
In [5]:
1  # Loop through the data frame and make predictions using auto ARIMA
2  for i in range(len(time_series) - T - h + 1):
3      # Define the training and testing data sets
4      train = time_series.iloc[i:i+T]
5      test = time_series.iloc[i+T:i+T+h]
6
7      # Fit the exponential smoothing model
8      model = SARIMAX(endog=train,
9                  exog=input_data.iloc[i:i+T].drop(columns = ['datetime', 'global']),
10                 order=order, seasonal_order=seasonal_order)
11     fit_model = model.fit(disp=False)
12
13     # Make predictions
14     preds = fit_model.forecast(steps=h,
15             exog=input_data.iloc[i+T:i+T+h].drop(columns = ['datetime', 'global']))
16
17     # Calculate percentage and absolute errors
18     perc_errors = np.abs(test-preds)/test
19     abs_errors = np.abs(test-preds)
20
21     # Store the percentage and absolute errors
22     perc_error_list.append(perc_errors)
23     abs_error_list.append(abs_errors)
24
25 perc_error_list = np.array(perc_error_list).reshape(len(time_series) - T - h + 1,h)
26 abs_error_list = np.array(abs_error_list).reshape(len(time_series) - T - h + 1,h)
```

Figure 3.23 Cell 5 of the Python Code for SARIMAX

Now let us turn to cell 5 depicted in Figure 3.23. The first seven lines are similar to the ARIMA case where we start by using the first T rows for training and test using the next row (recall $h = 1$). Then we again train from row 2 to row $T+1$, and test for $T+2$. We continue until the data set runs out. The train and test data selection are in lines 4 and 5. Lines 8, 9, and 10 define the model. Note that we use SARIMAX which was imported in cell 2. For the endogenous data we use the time-series called "train" while for the exogenous data we use all the columns in the data frame except the original time series (global) which we have already considered, and also drop the datetime column. The model also requires us to specify (p,d,q) and (P,D,Q) which we do in line 10.

Note that we can also use SARIMAX when there are no exogenous variables. For that all we need to do is write `exog = None` in line 9. In line 11 we fit the model. We can use `disp = True` if we want the fit to display whether it converged (this is because many times when there are exogenous variables, the optimization may not converge). Lines 14 and 15 together forecasts the next h values of the time-series. The exogenous data is the same as the one in line 9, except it ought to reflect the test data lines. The remaining lines are identical to that of ARIMA. Essentially we compare the predictions against the test.

Cells 6 and 7 inputs are identical to those in Figure 3.17 with the exception of the cell numbers in the figure being 5 and 6. The outputs however, are different and described in Table 3.5. The table provides the outputs as absolute percentage error (APE) and absolute error ratio (AER). Notice that the errors in all cases are lower than that with no exogenous variables. Although that is not a fair comparison as the parameters were different. We will do a thorough comparison later in this chapter.

Table 3.5
SARIMAX error metrics for APE and AER for a python code run with *h* = 1 and exogenous variables

Metrics	Mean	Median	75th perc.	90th perc.
Abs. percentage error	0.1395	0.0641	0.1471	0.3022
Abs. error ratio	0.0946	0.0596	0.1184	0.2151

However, it is worthwhile making this interesting observation. The mean and standard deviation of wind speeds *when nearby location was a good predictor* ranged 4.75–4.8 and 3.29–3.33, respectively. Also, for those good prediction situations, the wind direction had a mean from 185° to 189° while the standard deviation was 102° to 108° which points to a eastward wind that takes half hour from the nearby location to the main location.

That said, we move on to the next topic which is not a time-series prediction but more of a machine-learning approach.

3.3 FITTING RESIDUALS USING ENSEMBLE TREES

Although the section is titled "fitting residuals", here that means we will use a method that is built on a series of residual-based trees. However, we will consider hybrid models in the next chapter where we will use residuals besides using these ensemble trees. That said, here we begin by explaining the method of using ensemble of trees. In fact, the explaining is going to be very short with the understanding that details will be presented in the next chapter. By a tree, we mean a decision tree which is a binary tree. We take many decision trees (i.e., an ensemble) and make a prediction. In particular, we use a method called *Gradient Boosting* which uses a series of trees that are built by considering residuals, giving the nodes weights, and boosting them. There are other methods such as bagging (bootstrap aggregating) and random forest which are other forms of ensemble trees.

More details are beyond the scope but will appear briefly in the next chapter. For the moment it is worthwhile thinking of it as a regression (after all the method is called *gradient boosting regressor*). Like any regression model, our goal is to predict the global irradiance using features available at the time of making predictions. Now, we do not explicitly model as a time series but just take each row of the data set as an independent entry. However, we implicitly take into account the time series. In particular, we use the previous few observations as well as the same time for a few days in the past.

That said, let us get into the python code right away. The first cell as always is as before. Cells 2, 3, and 4 are in Figure 3.24. In cell-2 we are essentially importing the gradient boosting regressor library from the ensemble modeling module in scikit learn. In cell-3 we define the numerical values of the parameter similar to what

```
In [2]:
# Key imports for this code
from sklearn.ensemble import GradientBoostingRegressor
```

```
In [3]:
seasonal_periods = 32
# number of predictions to make
h = 1
# Define the length of each training set
T = 32*7
```

```
In [4]:
1  # Obtain solar irradiance data as a data frame
2  input_data = pd.read_csv('Oct2016_no_night_w_exog_solarTAC.csv', parse_dates = ['datetime'])
3  input_data['hour'] = pd.to_datetime(input_data['datetime'], format='%H:%M').dt.hour
4  input_data['minute'] = pd.to_datetime(input_data['datetime'], format='%H:%M').dt.minute
5  input_data['lag1'] = input_data['global'].shift(periods = 1)
6  input_data['lag2'] = input_data['global'].shift(periods = 2)
7  input_data['lag3'] = input_data['global'].shift(periods = 3)
8  input_data['lags'] = input_data['global'].shift(periods = seasonal_periods)
9  input_data['lag2s'] = input_data['global'].shift(periods = 2*seasonal_periods)
10 input_data.dropna(inplace = True)
11 input_data.head()
```

Out[4]:

	datetime	global	air_temp_k	global_exog	air_temp_k_exog	hour	minute	lag1	lag2	lag3	lags	lag2s
64	2016-10-03 08:00:00	399.37	293.42	280.44	291.24	8	0	293.65	289.96	495.45	332.14	392.13
65	2016-10-03 08:15:00	448.24	294.21	328.17	291.60	8	15	399.37	293.65	289.96	452.03	439.06
66	2016-10-03 08:30:00	495.50	294.43	373.78	291.99	8	30	448.24	399.37	293.65	480.37	483.33
67	2016-10-03 08:45:00	538.27	294.90	421.39	292.35	8	45	495.50	448.24	399.37	397.29	525.62
68	2016-10-03 09:00:00	577.93	295.36	465.79	292.58	9	0	538.27	495.50	448.24	571.07	566.48

Figure 3.24 Cells 2, 3 and 4 of the Python Code for Gradient Boosting

we did for ES and ARIMA. In particular, we say that the seasonality is 32 periods because we use eight hours of a day of sunlight, and within that we have four samples each hour (making it a total of 32). Then, we make predictions $h = 1$ period away. Also, we train for 1 week, i.e., $T = 32 \times 7$.

Now, let us focus on cell-4 in Figure 3.24. We use the sample data set (as shown in line-1) as in SARIMAX which includes the exogenous variable corresponding to the solar irradiance from a location to the west. Then in lines 3 and 4, we tease out the hour and minute from the column `datetime` with the understanding that the hour and minute as a combination could be a feature. In lines 5, 6, and 7, we obtain a lag-1, lag-2, and lag-3 features. For example, say the time now is 11am, then the lag-1 is the irradiance at 10:45, lag-2 at 10:30, and lag-3 at 10:15am. This is akin to using the previous few values of the time-series in ES and ARIMA.

Likewise, lines 8 and 9 correspond to irradiance at the location, 1 and 2 seasons away. That means, if it is 11am now, then line 8 gives the 11am irradiance yesterday, and line 9 offers the 11am irradiance day before. Note that while creating the lags, the first few lines will have `NaN` values because the lag values are not available. Then, in line 10, we drop the `NaN` values and print out the header of the input file. Note that `global` is the value we wish to predict using all the other columns (except `datetime`) as features. The output is displayed in the bottom of the figure. Note that the column lag1 has 293.65 as the first line and it appears in the second and third lines of lag2 and lag3.

Unlike ES and ARIMA where we train a model for every horizon in the sliding window and make a prediction 15 minutes out, here in gradient boosting we train

```
In [5]:  1 # Creating training data (only using 1 week, no re-training)
         2 X_train = input_data.iloc[0:T].drop(columns = ['datetime', 'global'])
         3 y_train = input_data['global'].iloc[0:T]
         4 X_train.head()
         5
         6 # defining the model and parameters
         7 gb = GradientBoostingRegressor(n_estimators = 250, max_depth = 4, min_samples_leaf = 3)
         8
         9 # Asking the model to fit the training data
        10 gb = gb.fit(X_train, y_train)
        11
        12 # Asking what the importance of features
        13 gb.feature_importances_
```

Out[5]:

	air_temp_k	global_exog	air_temp_k_exog	hour	minute	lag1	lag2	lag3	lags	lag2s
64	293.42	280.44	291.24	8	0	293.65	289.96	495.45	332.14	392.13
65	294.21	328.17	291.60	8	15	399.37	293.65	289.96	452.03	439.06
66	294.43	373.78	291.99	8	30	448.24	399.37	293.65	480.37	483.33
67	294.90	421.39	292.35	8	45	495.50	448.24	399.37	397.29	525.62
68	295.36	465.79	292.58	9	0	538.27	495.50	448.24	571.07	566.48

```
Out[5]: array([0.01410402, 0.03516298, 0.0068812 , 0.03201704, 0.00950856,
               0.81756694, 0.01996221, 0.03144582, 0.01427711, 0.01907412])
```

Figure 3.25 Cell 5 of the Python Code for Gradient Boosting

it just once and use it throughout the test data set. Refer to Figure 3.25. In lines 2 and 3, we set aside the training data as the first T lines of the original data frame. Like we said earlier, we do not consider the `datetime` column as a feature. Also, the predicted variable (or dependent variable) is `global`. In line 4 we ask to print the header of the training data features, and that is worthwhile looking at in the figure.

In line 7 of cell-5 shown in Figure 3.25, we invoke the gradient boosting regressor model. We choose to build 250 trees, and say that the depth of any of those trees can be a maximum of 4, and there must be at least 3 samples in each leaf of the tree. In line 10 we fit the train data. Also in line 10 we explicitly ask that the model is fit on the train data. Then in line 13 we ask the gradient boosting model to list what features are important. Note that the `lag1` feature with 0.81756694 as the most important one as it is the largest. So, if we know the current irradiance, we can predict the next one.

We use the test data to make predictions and then collect statistics for error metrics. This is done in cell-6 of the python code shown in Figure 3.26. Lines 2 and 3 initialize the vectors (or matrices for large h) to get percentage error and absolute error for each sample point. Then from lines 6 to 20 we go through a for-loop by considering each row of the test data (line 8) and predicting the irradiance (line 12). In line 9 we get the ground truth which is the observed irradiance. In lines 15 and 16 we compute the percentage error and the absolute error. Then in lines 19 and 20 we store the percentage and absolute error in the vector or matrix. Finally, in line 22 we reshape the data so there are h columns of predictions (here h is 1).

Cells 7 and 8 of the code are identical (in-fact verbatim) to those in cells 5 and 6 in Figure 3.17. Hence, we simply state the output of those two cells in Table 3.6. The first two lines of results correspond to the code described thus far by training for one week. In the last two lines we consider training for 2 weeks (hence $T = 32 \times 14$). This

In [6]:

```
# Initialize the lists to store the percentage and absolute errors
perc_error_list = []
abs_error_list = []

# Loop through the data frame and make predictions using Gradient Boosting
for i in range(len(input_data) - T - h + 1):
    # Define the testing data sets
    X_test = input_data.iloc[i+T:i+T+h].drop(columns = ['datetime', 'global'])
    y_test = input_data['global'].iloc[i+T:i+T+h]

    # Make predictions
    y_preds = gb.predict(X_test)

    # Calculate percentage and absolute errors
    perc_errors = np.abs(y_test-y_preds)/y_test
    abs_errors = np.abs(y_test-y_preds)

    # Store the percentage and absolute errors
    perc_error_list.append(perc_errors)
    abs_error_list.append(abs_errors)

perc_error_list = np.array(perc_error_list).reshape(len(input_data) - T - h + 1,h)
abs_error_list = np.array(abs_error_list).reshape(len(input_data) - T - h + 1,h)
```

Figure 3.26 Cell 6 of the Python Code for Gradient Boosting

Table 3.6

Gradient boosting error metrics for APE and AER for a python code run with h = 1 and different training lengths T

T	Metrics	Mean	Median	75th perc.	90th perc.
32×7	Abs. percentage error	0.1681	0.0654	0.1725	0.4483
32×7	Abs. error ratio	0.1012	0.0637	0.1341	0.2546
32×14	Abs. percentage error	0.1347	0.0508	0.1139	0.3000
32×14	Abs. error ratio	0.0798	0.0483	0.0926	0.2045

change is made by modifying line 5 of cell 3 shown in Figure 3.24. Notice from the results that by training for 2 weeks instead of 1, all the results have improved. Also, notice that the results are comparable to those of SARIMAX and ES even though we do not re-train the model before every test here unlike SARIMAX and ES. So this machine-learning based approach is extremely promising. It needs to be trained only once, it can take as much data as one can feed it, and it is much faster to run.

3.4 PERFORMANCE OF ALGORITHMS AGAINST PERSISTENT METHOD

The objective of this section is to impress upon the reader that there is no method that does well in terms of short-term solar forecasts for all conditions. We already saw some examples of that in the previous sections. Here we consider few other aspects and discuss some nuances that are important while making comparisons (restricting ourselves to $h = 1$, one step ahead forecast). The first of that nuance is that the methods must at the minimum be better than the so-called *persistent method*. This is a naive approach where the forecast for the next observation is simply the current observation, in other words we assume the atmospheric and solar conditions will persist (hence the name). Mathematically, if y_t is the observed irradiance at the tth interval, then the persistent method would forecast $\hat{y}_{t+h}$ as y_t. In other words, according to the persistent method, for all t,

$$\hat{y}_{t+h|t} = y_t.$$

As a first set of experiments, we compare the performance of various forecasting methods in different locations. The numerical values are in Table 3.7. Two of the three locations are something that we have considered before in Table 3.4, namely, Aurora, CO and Golden, CO. In addition, we also consider a location near Los Angeles, CA. The site is called SOLRMAP located in Loyola Marymount University (coded RSR). The data is from NREL and credit for the data source is to Andreas and Wilcox [4]. The Latitude is 33.966674° North while the Longitude is 118.4226° West. The elevation is 27 meters above mean sea level. We consider six forecasting methods: exponential smoothing (details in Section 3.1); GB w/o ES feature (this is identical to the model in Section 3.3); GB with ES feature which adds the ES prediction as a feature; Persistent fcst. (this is the aforementioned persistent forecast); SARIMAX via auto (the SARIMAX method in Section 3.2 where the parameters are selected by auto-arima); and SARIMAX auto corr (the SARIMAX method in Section 3.2 where the parameters are selected by considering auto-correlation plots).

Like before, we consider both APE (absolute percentage error) and AER (absolute error ratio). Which one to choose is dependent on the use case. However, more-or-less they present similar relative performance. As before, here too we consider four statistics mean, median, 75th percentile, and 90th percentile. It would be worthwhile to go over Table 3.7 carefully once before continuing reading. From the table, it appears like some methods cannot outperform the naive persistent forecast for some metrics (specifically the mean which is a popular metric). However, for the median case, the persistent forecast is mostly the worst. Likewise, for the median metric in the three sites considered, SARIMAX with auto corr performs the best. However, that method is one of the worst when it comes to 75th and 90th percentiles. So in a majority of predictions, SARIMAX with auto corr performs best. But in a minority of cases, it performs one of the worst.

It is disappointing to see the two GB methods not performing as well as they did in Section 3.3. However, that could be attributed to various factors. One of them is that the GB methods used `n_estimators = 100`, `max_depth = 5`, and `min_samples_leaf = 2` which are arbitrarily chosen. Ideally one would use a grid search to get the best parameters (not just the above but other default values as well). Further, GB is trained only once (i.e., the first week and using that predictions are made for the rest of the month) while ES and SARIMAX methods are re-trained for every prediction. In the forthcoming tables, this effect would become clearer. Another aspect to note is that the results for Aurora, CO which we saw in the previous sections are different. The reason is that here we used the month of September (in 2016) resulting in 10 hours of light (as opposed to 8 hours in the previous sections).

Similar to Aurora, the Golden CO location also used the same September 2016 data with 10 hours of light. However, the Los Angeles location used April 2015 data but with 10 hours as well. In every metric, the Los Angeles data produces lesser errors than the two CO locations. Note that unlike in Section 3.2, here we do not consider nearby locations' conditions as they are not available for all 3 sites. The (p,d,q) and (P,D,Q) parameters used for SARIMAX via auto were $(1,0,0)$ and $(1,0,2)$ for Aurora, $(2,1,1)$ and $(0,0,0)$ for Golden, and $(2,1,0)$ and $(0,0,2)$ for Los Angeles. The corresponding parameters used for SARIMAX auto corr were $(2,2,0)$ and $(0,0,1)$ for Aurora, Golden, and Los Angeles. Next we move to a different table.

Now, let us focus our attention to Table 3.8. Here we only use the Aurora, CO data but consider three time-intervals for aggregating irradiance: 5 minutes, 15 minutes, and 60 minutes. Table 3.7 only considered 15-minute granularities. The reason the results for 15-minute interval for Aurora do not match between the two tables is that for Table 3.8, we begin training only from the third day and test for two fewer days than what was done to get Table 3.7 results. The main reason for that is in Table 3.7, the GB method was trained for only 5 days, while in 3.8 it was trained for 7 days (however we drop the first two days as the lag features are unavailable).

There are a few other changes that are worth noting in Table 3.8. In the 5-minute granularity case, we only test for 1.5 weeks as opposed to the 3 weeks of testing in the other two granularities. Also, we removed the auto-ARIMA-based SARIMAX as in the 60-minute interval case the method did not produce any results (and threw errors due to non-convergence). For the SARIMAX with autocorrelation-based parameters, we used (p,d,q) and (P,D,Q) parameters as $(2,1,2)$ and $(0,0,0)$ for 5 min., $(2,2,0)$ and $(0,0,1)$ for 15 min., and $(3,2,0)$ and $(0,1,2)$ for 60 min. granularities.

There are three GB-based methods. The first two GB w/o ES feature and GB with ES feature are the same as in Table 3.7 where they correspond to the absence and presence respectively of the ES predictions. The new method here is GB with retraining. Here, similar to ES and SARIMAX, the model is trained for 1 week and the window is moved to the next time and the model is retrained. Given that background for generating the data for the experiments, next we discuss results of Table 3.8.

From a median prediction standpoint, for the 5 minute interval, the persistent forecast is the best. The reason is that the interval is so small that the irradiation remains practically the same. However for the other time intervals, persistent forecast does not yield as strong a result. In the 60-minute case, notice that exponential smoothing performs the strongest. It is also worthwhile noticing how much better GB with retraining performs. In the 15 minute granularity case, notice that exponential smoothing outperforms SARIMAX except in the median case.

As a final table we consider the training interval. This aspect was mentioned in the gradient boosting case in Section 3.3. There we said that if we have more training, GB performs better. In that light, in Table 3.9 we consider three train lengths: 4 days, 7 days, 14 days. We try the same 6 methods as in Table 3.8, hence it is not described here (the reader is encouraged to read the previous two or three paragraphs). An important feature is that the testing data in all three cases is exactly the same. It is

Table 3.7
Error metrics for various methods in multiple cities for 15 min. intervals

Location	Method	Metric	Mean	Median	75th perc.	90th perc.
Aurora, CO	Exp. Smoothing	APE	0.1362	0.0569	0.1471	0.3386
		AER	0.0931	0.0522	0.1163	0.2191
	GB w/o ES feature	APE	0.2270	0.0520	0.1997	0.5475
		AER	0.1032	0.0505	0.1385	0.2865
	GB with ES feature	APE	0.2285	0.0572	0.1976	0.5540
		AER	0.1074	0.0599	0.1390	0.2805
	Persistent fcst.	APE	0.1559	0.0769	0.1931	0.3678
		AER	0.1008	0.0700	0.0961	0.2461
	SARIMAX via auto	APE	0.1333	0.0564	0.1537	0.3149
		AER	0.0911	0.0546	0.0963	0.2020
	SARIMAX auto corr	APE	0.1471	0.0198	0.1704	0.4450
		AER	0.0920	0.0217	0.1177	0.2598
Golden, CO	Exp. Smoothing	APE	0.2199	0.0925	0.2072	0.4368
		AER	0.1318	0.0836	0.1688	0.3130
	GB w/o ES feature	APE	0.3069	0.0950	0.3017	0.6741
		AER	0.1468	0.0893	0.1976	0.3647
	GB with ES feature	APE	0.3300	0.1015	0.3039	0.7218
		AER	0.1518	0.0933	0.1955	0.3789
	Persistent fcst.	APE	0.2293	0.1042	0.2438	0.4608
		AER	0.1325	0.0858	0.1493	0.3288
	SARIMAX via auto	APE	0.2321	0.1146	0.2390	0.4750
		AER	0.1371	0.0966	0.1460	0.3110
	SARIMAX auto corr	APE	0.2300	0.0427	0.2827	0.6220
		AER	0.1289	0.0398	0.1686	0.3859
Los Angeles	Exp. Smoothing	APE	0.1039	0.0445	0.1242	0.2540
		AER	0.0755	0.0432	0.0988	0.1812
	GB w/o ES feature	APE	0.1197	0.0625	0.1560	0.2701
		AER	0.0965	0.0581	0.1420	0.2404
	GB with ES feature	APE	0.1130	0.0589	0.1421	0.2737
		AER	0.0872	0.0559	0.1176	0.2065
	Persistent fcst.	APE	0.1067	0.0620	0.1201	0.2422
		AER	0.0784	0.0616	0.0824	0.1592
	SARIMAX via auto	APE	0.0980	0.0388	0.1005	0.2564
		AER	0.0672	0.0355	0.0750	0.1638
	SARIMAX auto corr	APE	0.0988	0.0141	0.0990	0.3053
		AER	0.0650	0.0140	0.0740	0.1884

the last 14 days of the month of September in 2016. Again, the data is from Aurora. Prior to the start of the test date, we consider 4, 7, and 14 days for testing.

In Table 3.9, from a median performance standpoint, GB with ES and GB with retraining perform strongly. Thus, a method that uses retraining as well as ES may perform even better. A curious note is that it is not necessary that having more training days produce better results. The main reason for that is the trend is not accounted

Table 3.8
Error metrics for various methods under multiple time intervals in Aurora

Interval	Method	Metric	Mean	Median	75th perc.	90th perc.
5 min.	Exp. Smoothing	APE	0.1548	0.0602	0.1759	0.4090
		AER	0.1150	0.0590	0.1389	0.2803
	GB w/o ES feature	APE	0.2791	0.0795	0.2946	0.7160
		AER	0.1379	0.0715	0.2041	0.3525
	GB with ES feature	APE	0.3036	0.0916	0.2957	0.7280
		AER	0.1447	0.0802	0.2158	0.3640
	GB with retraining	APE	0.2133	0.0738	0.2707	0.5247
		AER	0.1289	0.0553	0.1707	0.3532
	Persistent fcst.	APE	0.1810	0.0595	0.2397	0.4409
		AER	0.1233	0.0362	0.1597	0.3527
	SARIMAX auto corr	APE	0.1525	0.0553	0.1522	0.4093
		AER	0.0991	0.0430	0.1033	0.2644
15 min.	Exp. Smoothing	APE	0.1415	0.0559	0.1518	0.3641
		AER	0.0945	0.0511	0.1186	0.2336
	GB w/o ES feature	APE	0.1930	0.0611	0.2046	0.4996
		AER	0.1024	0.0544	0.1359	0.2582
	GB with ES feature	APE	0.1995	0.0562	0.2126	0.4828
		AER	0.1024	0.0564	0.1263	0.2554
	GB with retraining	APE	0.1596	0.0481	0.1761	0.4185
		AER	0.0923	0.0434	0.1186	0.2556
	Persistent fcst.	APE	0.1625	0.0793	0.2070	0.3903
		AER	0.1037	0.0707	0.0969	0.2579
	SARIMAX auto corr	APE	0.1565	0.0240	0.1825	0.4727
		AER	0.0975	0.0230	0.1249	0.2733
60 min.	Exp. Smoothing	APE	0.2002	0.0843	0.2160	0.4668
		AER	0.1286	0.0851	0.1951	0.2982
	GB w/o ES feature	APE	0.3261	0.1339	0.3337	0.8669
		AER	0.1767	0.1255	0.2509	0.4011
	GB with ES feature	APE	0.2919	0.1214	0.2879	0.7239
		AER	0.1626	0.1140	0.2334	0.3674
	GB with retraining	APE	0.2484	0.0956	0.2746	0.6875
		AER	0.1453	0.0820	0.2235	0.3688
	Persistent fcst.	APE	0.3606	0.2448	0.4153	0.8313
		AER	0.2369	0.2143	0.3468	0.4174
	SARIMAX auto corr	APE	0.2397	0.1022	0.2422	0.6223
		AER	0.1438	0.1014	0.1920	0.3485

for. Two weeks are long enough. Perhaps using a year's worth data with a feature of date and month could improve the method. Notice that the persistent forecast produces the same result no matter the train length. This is because it uses just the previous value and not all the train data. Also, the test data is identical in all cases.

Table 3.9
Error metrics for various methods under different training lengths in Aurora

Train length	Method	Metric	Mean	Median	75th perc.	90th perc.
4 days	Exp. Smoothing	APE	0.1299	0.0612	0.1507	0.3325
		AER	0.0864	0.0453	0.1227	0.2108
	GB w/o ES feature	APE	0.3284	0.0672	0.3296	0.9112
		AER	0.1222	0.0534	0.1894	0.3498
	GB with ES feature	APE	0.2218	0.0407	0.2713	0.6958
		AER	0.0943	0.0290	0.1361	0.2394
	GB with retraining	APE	0.2815	0.0551	0.2724	0.8928
		AER	0.1124	0.0378	0.1681	0.2773
	Persistent fcst.	APE	0.1519	0.0776	0.2042	0.3406
		AER	0.0930	0.0681	0.0949	0.2128
	SARIMAX auto corr	APE	0.1323	0.0463	0.1702	0.3324
		AER	0.0849	0.0431	0.0916	0.1997
7 days	Exp. Smoothing	APE	0.1188	0.0489	0.1243	0.3210
		AER	0.0781	0.0429	0.0909	0.1739
	GB w/o ES feature	APE	0.1098	0.0276	0.0913	0.3900
		AER	0.0637	0.0252	0.0642	0.1742
	GB with ES feature	APE	0.1253	0.0340	0.1225	0.3335
		AER	0.0690	0.0375	0.0777	0.1935
	GB with retraining	APE	0.0897	0.0240	0.0844	0.2413
		AER	0.0520	0.0217	0.0535	0.1287
	Persistent fcst.	APE	0.1519	0.0776	0.2042	0.3406
		AER	0.0930	0.0681	0.0949	0.2128
	SARIMAX auto corr	APE	0.1399	0.0686	0.1830	0.3381
		AER	0.0893	0.0624	0.0985	0.1864
14 days	Exp. Smoothing	APE	0.1210	0.0446	0.1227	0.3280
		AER	0.0744	0.0397	0.0859	0.1793
	GB w/o ES feature	APE	0.1660	0.0528	0.1959	0.4415
		AER	0.0915	0.0459	0.1064	0.2411
	GB with ES feature	APE	0.1255	0.0323	0.1210	0.3776
		AER	0.0689	0.0267	0.0780	0.1700
	GB with retraining	APE	0.1418	0.0455	0.1565	0.3865
		AER	0.0819	0.0377	0.1018	0.2104
	Persistent fcst.	APE	0.1519	0.0776	0.2042	0.3406
		AER	0.0930	0.0681	0.0949	0.2128
	SARIMAX auto corr	APE	0.1343	0.0661	0.1647	0.3091
		AER	0.0836	0.0592	0.0904	0.1702

In summary, we can state that the quality of the results can vary from location to location, from one method to another, from one set of parameters to another, and from what metric is chosen. Care must be taken to select the appropriate metric. For some cases median may be the best while mean is the best for others. In some applications people may prefer to have the worst case as small as possible. Also, the quality of a method depends on the parameters used, the time of the year, the time interval, and the length of the training set. Another aspect is that during some intervals one method may perform better than others. If they are carefully investigated, it would be possible to provide much more accurate and meaningful forecasts.

4 Day-Ahead Solar Forecasts

The aim of this chapter is to describe a few approaches for *day-ahead* forecasts of solar irradiance. In many ways this chapter is similar to the previous chapter on *short-term* forecasts. For that reason, things would not be re-stated in this chapter and it would be useful to read up on those from the previous chapter. In other words, skipping the previous chapter and starting to read this chapter is not advisable. At least a cursory glance or speed-reading of the previous chapter would be necessary. Having said that, there are also significant differences between the treatment of short-term and day-ahead forecasts. However, it is critical to understand that the differences are not always because things are done a certain way or certain methods are used in short-term versus day-ahead forecasts. Before reading ahead, it would be useful to reflect on that.

In this chapter we will continue to use exponential smoothing and SARIMAX that were described in the previous chapter. However, in this chapter we will not go over the details and the reader is encouraged to look at the previous chapter. Further, we use XGBoost and describe it in detail. However, one could simply use Gradient Boosting, like it was done in the previous chapter. That said, let us jump into *day ahead* forecasts of solar irradiance. The main motivating reason for day-ahead forecasts is that it helps in planning especially for utilities that source power from multiple modes. For that reason, the day-ahead price is often cheaper than real-time price of power. So it is extremely important to be able to forecast solar irradiance a day ahead. In this book we will stick with 1-hour granularity, although it would easily be possible to consider others.

We saw in the previous chapter that it is difficult to predict solar irradiance accurately even 15 minutes out. Then, what is the hope for day-ahead forecasts? That concern is valid and hence there is a need for additional features. One of them is the weather. A cloudy hour would mean the solar irradiance is lower than if the sky would be clear that hour. Also, weather forecasts a day in advance is extremely accurate. In fact days that are overcast are relatively easy to predict. However, exponential smoothing as a method cannot handle exogenous variables. For that reason we would use the predictions of exponential smoothing inside our SARIMAX model as an exogenous variable, besides the weather.

We focus on developing a *point forecast* for the amount of power a solar PV panel will generate averaged over an hour. This is with the understanding that an interval forecast would be generated in a manner similar to what was done in the previous chapter. We assume that forecasts need to be made a day in advance, for example, to predict the average amount of solar irradiance from 10-11am tomorrow, only information available is the hourly average irradiance before today's 11am. Similar to

DOI: 10.1201/9781003403500-4

the short-term forecasts, here too we only use parts of a day (and delete remaining data), for example, 8 hours of solar in a day (8am-4pm). To obtain hourly average, we take the 60 minute-by-minute data and compute their mean. So, if we used 8 hours of data each day, there would be 8 samples each day: 8am (corresponding to 8:00-8:59), 9am, ..., 3pm (i.e., 3:00-3:59pm). In the case where we have 8 samples each day, we will forecast 8 periods ahead (in short-term, we only used 1 period ahead as the basic case).

4.1 USE OF WEATHER PREDICTIONS

Recall from the previous chapter the forecasts made for several periods later are not accurate. Purely using irradiance will not cut it. Also from a physical standpoint, the irradiance depends on the time of the day as well as whether there is cloud cover. Clearly, cloud conditions vary a lot, so knowing the irradiance now may have little or no impact tomorrow. However, weather forecasting is extremely accurate. It would be possible to obtain fairly accurately the weather conditions and temperature. Our goal is to use that information to make better forecasts. These days it is possible to obtain day-ahead weather for any location, if we take out our phones, we can see the weather at our location. We might as well use it.

Talking of phones, in practice it would be possible to simply use the weather forecast on the phones as it gives it for a fairly compact region. In fact, the NOAA (National Oceanic and Atmospheric Administration) and many weather-related sites provide forecasts. As it turns out, many also have the ability to get weather forecasts for historic dates so that models can be built. However, for the purposes of this chapter we used data from Kaggle which has datasets for many cities with the understanding that the data may be from a location far from the solar PV panel where we seek to predict the irradiance. Thus in practice we recommend using forecasts of locations as close to the solar PV panels as possible.

Although it may be possible to use a variety of metrics, we mainly use two: ambient conditions (which we simply refer to as weather) and temperature. Ambient conditions are categorical variables such as: clear sky; scattered clouds; partly cloudy; thunderstorm; overcast; foggy; rainy; snowy; and so on. We begin by performing a simple mapping by using 2 for clear-sky (expecting to see the highest irradiance), 1 for partly cloudy conditions, and 0 for what we believe is likely to be overcast. For such a mapping, if we were to obtain the correlation between the irradiance and 0-1-2 ambient conditions, we do not observe a very high correlation for a year's worth of data. This is because the irradiance significantly changes over the year. However, if we use a single season of 3 months, the correlation is higher.

Finally, we scale the irradiance (a topic we will consider in the next section) by dividing the observed irradiance by what is possible if the sky was clear in that hour. This way this number is a continuum between 0 and 1 (assuming there is no diffusion). If we took the correlation between the scaled irradiance (process will be explained in the next section), we observe a much higher correlation with the ambient conditions than when we had unscaled data or annual data. The reverse is true for temperature, i.e., for the scaled data temperature is independent of irradiance.

But there is a strong correlation for the all-year unscaled data. This is because the temperature has a diurnal pattern similar to the irradiance, after all, both are caused by the sun.

For temperature we would perform a simple encoding by converting to degrees Kelvin (although this is not necessary, it is useful because the physics-based models between temperature and pressure use temperature in degrees Kelvin). Unlike what we did for correlations above, in our forecasts we will not map ambient conditions to a numerical value but use what is called one-hot encoding. In other words, each of the categories would take values of 0 or 1 if that is false or true. So a clear sky hour will have a 1 for the condition ‘clear sky’ and a 0 for all the others such as scattered clouds, partly cloudy, etc. One of the downsides is that there could be a vast number of different ambient conditions. In that case we restrict to about 8-10 popular types and club the rest into a category called ‘other’.

For the years 2016-17 in Denver, CO the Kaggle data showed the following ambient conditions: ‘drizzle’, ‘proximity shower rain’, ‘heavy snow’, ‘light intensity drizzle’, ‘moderate rain’, ‘heavy intensity rain’, ‘proximity thunderstorm’, ‘snow’, ‘dust’, ‘light rain and snow’, ‘heavy intensity drizzle’, ‘thunderstorm with light rain’, ‘thunderstorm with rain’, ‘thunderstorm with light drizzle’, ‘thunderstorm with heavy drizzle’, ‘proximity thunderstorm with rain’, ‘smoke’, and ‘thunderstorm’. However they were relatively rare, and it would make sense to combine them all into a category called ‘other’. The main reason for doing that is that with so many categorical variables, the model would lose fidelity.

Then in Denver, CO, the following are the most common weather conditions in 2016–17 and merit their own one-hot encoding. These are: ‘fog’, ‘light rain’, ‘light snow’, ‘overcast clouds’, ‘haze’, ‘mist’, ‘broken clouds’, ‘scattered clouds’, ‘few clouds’, and ‘sky is clear’. The last one is by far the most common and thus making a strong case to use solar panels. It is not difficult to predict the irradiance when the sky is clear. In fact even for overcast conditions, it is not difficult. The challenge comes under partly cloudy conditions because that can result in a lot of errors as it is not apparent if that would result in a shadow on the solar PV panel. These uncertainties are best left to some of the machine learning models that we will use for our forecasts.

To see how to incorporate such categorical variables that describe weather, let us go through a code for gradient boosting. In the process our goal is also to explain various nuances to be considered in day-ahead forecasting as well as the gradient boosting method. It is worthwhile noting that the code itself is not optimized to solve this problem. As we describe the code, we will also discuss various ways in which the approach could be optimized. In fact, we leave out a key feature in this presentation which is the forecast from exponential smoothing. Adding it would improve the accuracy of the results, and would be an important first step. The speed of SARIMAX and its accuracy are not high enough to be considered but that would be an option.

With that motivation and caveats aside, we now describe the Python code. Like all the Python notebook codes in this book, here too we will use a first cell as identical

In [2]:
```
# Key imports for this code
from sklearn.ensemble import GradientBoostingRegressor
```

In [3]:
```
seasonal_periods = 8
# fetch data
prev_year = pd.read_csv('solarTAC_hourly_2016_4_months_unscaled.csv', parse_dates = ['datetime'])
solar_data = pd.read_csv('solarTAC_hourly_2017_4_months_unscaled.csv', parse_dates = ['datetime'])
prev_year.head()
solar_data.head()
```

Out[3]:

	datetime	hour	global	Denver_temp	Denver_weather
0	2016-09-01 08:00:00	8	550.69	287.84	sky is clear
1	2016-09-01 09:00:00	9	717.42	287.65	sky is clear
2	2016-09-01 10:00:00	10	832.57	287.05	sky is clear
3	2016-09-01 11:00:00	11	895.11	286.64	sky is clear
4	2016-09-01 12:00:00	12	893.44	285.95	sky is clear

Out[3]:

	datetime	hour	global	Denver_temp	Denver_weather
0	2017-08-01 08:00:00	8	625.77	290.59	sky is clear
1	2017-08-01 09:00:00	9	777.04	289.81	sky is clear
2	2017-08-01 10:00:00	10	918.05	288.93	sky is clear
3	2017-08-01 11:00:00	11	977.27	288.25	sky is clear
4	2017-08-01 12:00:00	12	986.46	287.36	sky is clear

Figure 4.1 Cells 2 and 3 of the Python Code for Day Ahead Forecasting with Gradient Boosting

to that in Figure 1.6. The second and third cells are shown in Figure 4.1. Notice that the second cell is identical to that in Figure 3.24 used for short-term forecasts. Since we are using the same gradient boosting regressor here as well, Cell 2 remains unchanged. The big change is in Cell 3. Notice from Figure 4.1 that we are reading two CSV files. One is three months of data from 2016 and the other from 2017 is for four months. Our objective is to use the 2016 data to train and test on the 2017 data without retraining the model.

We continue with cell 3 in Figure 4.1. From the output of the header of 5 rows for both the 2016 and 2017 data, notice a few things. One is that the columns of `datetime` and `hour` show that the data is hourly, as described earlier in this chapter. Also, we use 8 hours of sunlight in a day, and hence the number of seasonal periods is 8 (as written in the first line of the cell). Also notice from the column `datetime` that the three months of 2016 starts from September, while the four months in 2017 start from August. So there is no training data for the month of August, although we predict for August. The dependent variable (i.e., the one we will forecast) is the column `global`.

The last two columns are the temperature in Denver (and it is a forecast) in degrees Kelvin and the weather in Denver (also a forecast). The categorical variable of Denver weather takes a single value of `sky is clear` in all the header lines. Also note that the predictions for temperature and weather in this data set correspond to the city of Denver, although the data is from Solar TAC which as described in the previous chapter is in the suburbs. In the ideal situation, one would obtain forecasts of temperature and weather for a location close to the solar panel site. The temperature ends up not being an important feature, but the weather is. However, weather is a categorical variable and care needs to be taken to model it.

In [4]:
```
input_data = pd.get_dummies(solar_data, drop_first=True)
input_data.head()
input_data.columns
```

Out[4]:

	datetime	hour	global	Denver_temp	Denver_weather_few clouds	Denver_weather_fog	Denver_weather_haze	Denver_weather_light rain	Denver_weather_light snow	Denver_
0	2017-08-01 08:00:00	8	625.77	290.59	0	0	0	0	0	
1	2017-08-01 09:00:00	9	777.04	289.81	0	0	0	0	0	
2	2017-08-01 10:00:00	10	918.05	288.93	0	0	0	0	0	
3	2017-08-01 11:00:00	11	977.27	288.25	0	0	0	0	0	
4	2017-08-01 12:00:00	12	986.46	287.36	0	0	0	0	0	

Out[4]:
```
Index(['datetime', 'hour', 'global', 'Denver_temp',
       'Denver_weather_few clouds', 'Denver_weather_fog',
       'Denver_weather_haze', 'Denver_weather_light rain',
       'Denver_weather_light snow', 'Denver_weather_mist',
       'Denver_weather_other', 'Denver_weather_overcast clouds',
       'Denver_weather_scattered clouds', 'Denver_weather_sky is clear'],
      dtype='object')
```

Figure 4.2 Cell 4 of the Python Code for Day Ahead Forecasting with Gradient Boosting

See cell 4 of the code in Figure 4.2. This is where we convert the categorical variables to binary variables. For each category, a dummy variable is created which takes the value of zero or 1. Since the number of categories is large in the Denver weather example, we consider another categorical variable: color. Say it takes three values red, blue, and green. Then the method would drop the column called "color" and create three new columns called "color˙red", "color˙blue", and "color˙green". Then, say a row of original data has a color value of blue, then the column "color˙red" would take the value 0, "color˙blue" would be 1, and "color˙green" would be zero.

This method of forming dummy variables is done using the first line in Figure 4.2 where use the command "get˙dummies". There are other methods, and the most popular of them is "one hot encoding". There are a few more lines of code in that case and hence not used. But that would also be more appropriate here, as we would see in one of the next cells. In the first line there is also a command `drop_first = True`. What that does is drops the first category. So in the red, blue, and green example, there would be no column called "color˙red" as we know if blue and green are zero, then it must be red. The main reason for that is that while using linear regression (which is not the case for gradient boosting but is for other time series models), it would result in linearly dependent columns with the intercept. As a result the inversion and decomposition would fail.

The result of the second line of the input to cell 4 produces the data frame header. Since the number of columns is large, only a few are displayed. Recall from Figure 4.1 that all the header rows' value for "`Denver_weather`" were `sky is clear`. So only the column `Denver_weather_sky is clear` will have ones while the others will be zero. That column corresponding to clear sky is not displayed. However, notice in the next cell where the code is asked to display the names of all the columns. There the output shows the clear sky column is in fact the last one. Also notice that all

In [5]:

```
1 input_data['lags'] = input_data['global'].shift(periods = seasonal_periods)
2 input_data['lags-1'] = input_data['global'].shift(periods = seasonal_periods+1)
3 input_data['lag2s'] = input_data['global'].shift(periods = 2*seasonal_periods)
4 input_data.dropna(inplace = True) # This will drop the first two full days
5 input_data.head()
```

Out[5]:

er_weather_light snow	Denver_weather_mist	Denver_weather_other	Denver_weather_overcast clouds	Denver_weather_scattered clouds	Denver_weather_sky is clear	lags	lags-1	lag2s
0	0	1	0	0	0	621.58	689.84	625.77
0	0	1	0	0	0	793.58	621.58	777.04
0	0	1	0	0	0	902.04	793.58	918.05
0	0	0	0	0	0	915.59	902.04	977.27
0	0	1	0	0	0	967.36	915.59	986.46

Figure 4.3 Cell 5 of the Python Code for Day Ahead Forecasting with Gradient Boosting

the main weather types have been categorized and the less frequent ones are clubbed into "other". This itself may be too many categories and it may be wiser to combine some of the features.

Next consider Cell 5 in Figure 4.3. Before describing the lines of the code input, let us focus on the output. For purposes of illustration, the rightmost columns of the data frame are presented. The reason the `Denver_weather_sky is clear` column does not have ones but `Denver_weather_other` has many ones is that we are actually not viewing the same header rows as in Cell 4. We will explain that subsequently. Also note that the last columns were not there in the original data. In fact, we create those columns in the first three lines of the inputs in Cell 5 as shown in Figure 4.3. The first line in Cell 5 essentially shifts the data in column `'global'` by 8 positions (where 8 is the number of seasonal periods, i.e., hours of sunshine modeled in a day).

The syntax for making that shift is described in line 1 of the input to Cell 5. The result is stored in a new column called `'lags'`. The reason we shift and create a new column is so that when we have to predict solar irradiance for a particular hour, we know the irradiance experienced the previous day at that same hour. This is essentially the day-ahead forecast. In other words, given the irradiance now, we ask what would it be after 8 (equal to seasonal periods) data points later. So, although we train the model using the previous year, the features used in prediction still correspond to the irradiance seen the same hour of the previous day. We will say more about that subsequently.

Next consider line 2 of the input to Cell 5. The syntax is very similar to line 1. Notice that here we shift by 9 (i.e., one more than the seasonal periods). The reason for that can be explained with an example. Say that we wish to forecast tomorrow's irradiance 10–11am. We use today's forecast for 10–11am (and anything after 10–11am today is not allowed). However, it could be that today 10–11am there is a passing cloud that resulted in a low irradiance, then we can use 9–10am to see if those numbers can be used as a feature. This new column is called `'lags-1'`. In a similar fashion, line 3 obtains a new column `'lag2s'` as it used 2 seasonal periods ahead. For the above example, that would mean 10–11am yesterday.

```
In [6]:  1 gb_train_data = pd.get_dummies(prev_year, drop_first=True)

In [7]:  1 gb_train_data['lags'] = gb_train_data['global'].shift(periods = seasonal_periods)
         2 gb_train_data['lags-1'] = gb_train_data['global'].shift(periods = seasonal_periods+1)
         3 gb_train_data['lag2s'] = gb_train_data['global'].shift(periods = 2*seasonal_periods)
         4 gb_train_data.dropna(inplace = True) # This will drop the first two full days

In [8]:  1 # Needed to add these two because the previous year did not have these
         2 input_data['Denver_weather_other'] = input_data['Denver_weather_other'] + input_data['Denver_weather_light snow'
         3 input_data.drop(columns = ['Denver_weather_light snow'], inplace = True)
```

Figure 4.4 Cells 6, 7, 8 of the Python Code for Day Ahead Forecasting with Gradient Boosting

The motivation for using 2 seasonal periods is that if today is overcast and tomorrow is forecasted to be sunny, then the approach would need some inputs to make the forecast, and the hope is that the previous day was sunny. In this way we are able to capture the day-to-day variability by considering enough samples. When we shift the irradiance column by 2 seasonal periods, it would result in `NaN` for the first 16 rows. If there is data from the previous 16 days, this can be averted. However, if it is not, we could simply drop those rows. That syntax is in line 4 of Cell 5. We drop the rows with `NaN`. One must be careful to check that there are no `NaN` values in other columns (otherwise, there is a way to use a subset of columns to check).

Next we focus our attention on Cells 6, 7, and 8 as shown in Figure 4.4. In Cell 6 we again create dummy variables but this time for the previous year's data (note that in Cell 4 shown in Figure 4.2, we used the forecast year). As described earlier, we use the previous year's data to train the gradient boosting model. For that reason the resulting data frame is called `gb_train_data`. Further, in Cell 7 in Figure 4.4 we create three new columns of the data frame by shifting the irradiance column by 8, 9, and 16, denoting 1 seasonal period, an hour more than 1 seasonal period, and 2 seasonal periods, respectively.

In Cell 8 of Figure 4.4, we need to make an adjustment. We notice that the Denver weather column did not have a category called "light snow" in the training data (previous) year, but it was there in the test data (current) year. If we were to use one-hot encoding, this situation may have been averted, as the code will not throw an error. However, it is unclear if it would cause other issues such as a wrong classification. So, as a work around we group "light snow" into the "other" category (see line 2 in Cell 8). Then we drop the column corresponding to "light snow" in line 3. This way both the training and the testing data have the same number of columns. In practice while running such a code in real time, one would not have this luxury, and guardrails to avoid these must be in place.

Now we have all the data to train the model and test it. Cells 9 and 10 shown in Figure 4.5 correspond to training and testing respectively. In Cell 9 we create a training set of features by collecting the columns in the training data frame in line 2. Notice that we drop the columns `datetime` and `global`. The important features of the former is captured in the "hour" column, while the latter forms the dependent variable in line 3. It is common to use X as the matrix of independent variables and y as a column of dependent variables, hence the name `X_train` and `y_train` for

```
In [9]:  1 # Creating training data using previous year
         2 X_train = gb_train_data.drop(columns = ['datetime', 'global'])
         3 y_train = gb_train_data['global']
         4 # defining the model and parameters
         5 gb = GradientBoostingRegressor(n_estimators = 100, max_depth = 5, min_samples_leaf = 2)
         6 # Asking the model to fit the training data
         7 gb = gb.fit(X_train, y_train)
         8 # Asking what the importance of features
         9 gb.feature_importances_

Out[9]: array([1.48849001e-02, 9.09069863e-02, 6.17966095e-03, 0.00000000e+00,
               6.06643711e-05, 2.51625288e-03, 7.29297865e-05, 1.50492656e-03,
               2.15730144e-03, 8.83362615e-04, 3.52957620e-02, 2.65253092e-01,
               4.07314443e-02, 5.39552717e-01])
```

```
In [10]:  1 # Initialize the lists to store the percentage and absolute errors
          2 gb_perc_error_list = []
          3 gb_abs_error_list = []
          4 # Define the testing data sets
          5 X_test = input_data.drop(columns = ['datetime', 'global'])
          6 y_test = input_data['global']
          7 # Make predictions
          8 y_preds = gb.predict(X_test)
          9 # Calculate percentage and absolute errors
         10 perc_errors = np.abs(y_test-y_preds)/y_test
         11 abs_errors = np.abs(y_test-y_preds)
         12 # Store the percentage and absolute errors
         13 gb_perc_error_list.append(perc_errors)
         14 gb_abs_error_list.append(abs_errors)
```

Figure 4.5 Cells 9 and 10 of the Python Code for Day Ahead Forecasting with Gradient Boosting

the features and ground truth columns respectively in lines 2 and 3 of Cell 9. In the forecasting (or testing) phase, if we are given the features, we must be able to make a prediction.

In line 5 of Cell 9 we define the model as a gradient boosting regressor. We select hyper-parameters `n_estimators`, `max_depth`, and `min_samples_leaf` as 100, 5, and 2, respectively. That means we choose to build 100 trees, and say that the depth of any of those trees can be a maximum of 5, and there must be at least 2 samples in each leaf of every tree. In a later section we will define these parameters, however, for now it is crucial to note that they would have to be selected carefully. Optimizing them is a critical part of model development, and that is not done in this code. We will discuss aspects of hyper-parameter tuning when we describe ensemble of trees subsequently.

In line 7 of Cell 9, we fit the training data. Notice that we ask it to fit using the train data defined in lines 2 and 3. This is for the previous year's data. Then in line 9 we ask the gradient boosting model to list what features are important. The order of features are the same as that in the columns. So `lag2s` and `lags` are the most important features as they are the largest two. That makes sense because if we wish to predict tomorrow's irradiance between 10am and 11am, the most important features are today's 10–11am and yesterday's 10–11am. However, it is surprising that yesterday's seems more important. But those values are also dependent on the other parameters. It is possible that the weather forecasts adversely affect the impact of today's forecast for tomorrow.

Next we move to Cell 10 which is used to test the model. As stated earlier, we test for 4 months of data from the year 2017. Lines 2 and 3 initialize the vector of

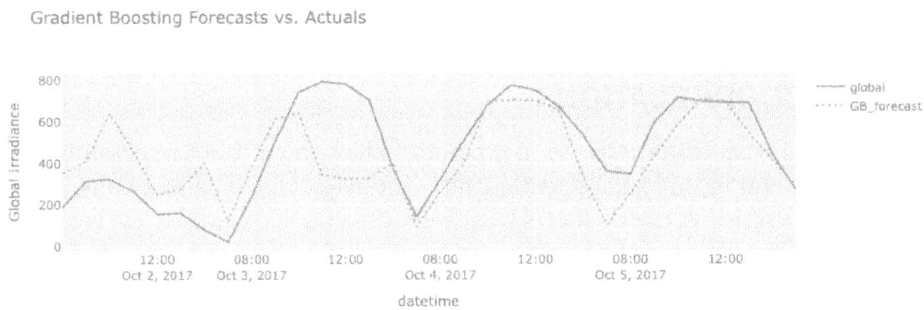

Figure 4.6 Comparing the Actual Irradiance against that forecasted by Gradient Boosting Trained with Previous Year Data

errors for each forecast, where line 2 is for absolute percentage error while line 3 for absolute error. In line 5 of Cell 10 we create the features used to make the forecasts in the test set. Notice the similarity between lines 5 and 6 of Cell 10 against lines 2 and 3 of Cell 9 respectively. The test set essentially mirrors the train set. However the array of ground truth values in the test set, `y_test` is purely for checking the quality of the forecasts and not used in any way in the forecasting model itself.

Line 8 in Cell 10 essentially uses the features in the test set and makes a prediction for each row. The predicted values are stored in an array `y_preds`. In lines 10 and 11 we obtain arrays of absolute percentage errors and absolute errors. The command `np.abs()` uses the library numpy to compute the absolute value. Python does a computation row by row, especially in line 10 to obtain the percentage error for each prediction. For the error notice that we compare the ground truth `y_test` against the forecasted values `y_preds`. Lines 13 and 14 in this code block are not adding extra value. However, if we chose to retrain the model after getting a few samples in 2017, this would be useful.

We do not present the remaining steps of the Python code, especially the results. We will do that in the last section of this chapter. That way we can compare against multiple methods. However, to give a flavor of the predictions, we take a visual approach as shown in Figure 4.6. As done in the previous chapter, here too we only consider the 8 hours of each day and not present the times not used. The line corresponding to `global` is the actual hourly average irradiance and the `GB_forecast` is the forecasted values. Four consecutive days are presented. Notice that the predictions are reasonable on the first, third, and fourth days.

Notice from Figure 4.6 that the second day's forecast on a fairly clear sky day appears to be poor. Upon looking carefully, it was found that the predicted weather was not clear sky, but some form of precipitation. Also, the values from the previous day were significantly low. Hence the model essentially forecasted the way it did. However, the weather forecasts were more reasonable on the other three days and the model ended up performing reasonably well. One can argue that this is not perfect especially for a day that appears to have clear skies. The model takes a fraction of a second to train and run, so one could easily incorporate more recent training, if needed.

4.2 AUTOCORRELATIONS

To understand what features to use to predict and how to go about forecasting a day ahead, it is useful to understand and quantify the sequential data that comes in the form of a time series. A metric that is frequently used for such an analysis is autocorrelation. We first need to explain what autocorrelation is, how the methods measure it, and how it (or other similar metrics related to it) can be used to make forecasts. In fact, the autocorrelation function can be used to determine which features to consider for the purposes of forecasts. Hence, it is an extremely useful metric or function, and thus worthwhile studying it carefully.

Let y_1, y_2, ..., y_T be the average solar irradiance corresponding to indices 1, 2, ..., T. This is the time-series of irradiance we have and we wish to see the autocorrelation of this time series. To explain autocorrelation, we first discuss correlation. Two random variables X and Z are correlated means that if one takes a high value, the other is expected to take a high value and vice versa. Mathematically, correlation is computed using the Pearson's coefficient of correlation which is

$$\rho_{X,Z} = \frac{\mathbb{E}[XZ] - \mathbb{E}[X]\mathbb{E}[Z]}{\sqrt{\mathbb{V}[X]}\,\sqrt{\mathbb{V}[Z]}} \tag{4.1}$$

where $\mathbb{E}$ is the expectation operator and $\mathbb{V}$ is the variance operator.

Equation (4.1) is a probabilistic expression and can be computed if we know the joint-distribution of X and Z. However, in practice, one typically estimates the Pearson's coefficient of correlation using samples of X and Z. Say, for example, that there are n samples of X and n corresponding samples of Z. We call those n samples as x_1, $x_2, \ldots, x_n$ of X and $z_1, z_2, \ldots, z_n$ of Z. Using those n samples, the Pearson's coefficient of correlation is estimated as

$$\hat{\rho}_{X,Z} = \frac{\sum_{i=1}^{n}(x_i - \bar{x})(z_i - \bar{z})}{\sqrt{\sum_{i=1}^{n}(x_i - \bar{x})^2}\,\sqrt{\sum_{i=1}^{n}(z_i - \bar{z})^2}} \tag{4.2}$$

where $\bar{x}$ is the mean of X given by $(x_1 + x_2 + \ldots + x_n)/n$ and $\bar{z}$ is the mean of Z given by $(z_1 + z_2 + \ldots + z_n)/n$.

Although it may not be too difficult to see that $\bar{x}$ is an estimate of $\mathbb{E}[X]$ and $\bar{x}$ is an estimate of $\mathbb{E}[Z]$, it may be a stretch to see how we went from Equation (4.1) to Equation (4.2). It can be explained using the facts that $\mathbb{E}[XZ] = \mathbb{E}[(X - \mathbb{E}[X])(Z - \mathbb{E}[Z])] + \mathbb{E}[X]\mathbb{E}[Z]$ and $\mathbb{V}[X] = \mathbb{E}[(X - \mathbb{E}[X])^2]$ that translate to the numerator and denominator, respectively. Every expectation operator is estimated by taking the average of the n samples.

Having explained correlation and how it is computed, now we turn our attention to autocorrelation. Suppose that we have a time-series. It is a collection of a sequence of random variables that represent the same process at different times. In particular, at time zero say that Y_t and Y_{t+k} be the random variables corresponding to the *t-th* and $t+k\text{-}th$ terms that will be observed in the future. An estimate of the correlation between Y_t and Y_{t+k} for some k and any t is called the autocorrelation. In some sense it is the same metric, hence the term "autocorrelation" (correlation with itself) but it is crucial to note that it is at different times.

In terms of the hourly solar irradiance, we explain it as follows. It may be that the irradiance between now and 3 hours from now may be correlated. That is because of clear sky or cloudy conditions that typically last for a few hours. Mathematically speaking this is similar to getting the correlation of Y_t and Y_{t+3} (note that $k=3$ due to the 3 hour gap). Now, we have T samples of irradiance $y_1, y_2, \ldots, y_T$. We define an arbitrary lag k (for some integer $k \geq 0$) for which we wish to determine the correlation between Y_t and Y_{t+k}, i.e., autocorrelation. Software packages would not allow k to be larger than $T/2$ or $T/3$.

Unlike the X and Z case we saw in Equation (4.1) where we got estimates of the expected values as $\bar{x}$ and $\bar{z}$, here the estimates of the expected values of Y_t and Y_{t+k} are both $\bar{y} = (y_1 + y_2 + \ldots + y_T)/T$. Further, the number of samples to estimate the covariance (numerator of Equation (4.1) is called covariance) is n pairs of samples for X and Z, however for Y_t and Y_{t+k}, we have $T-k$ pairs of samples (y_1 and y_{k+1}, y_2 and y_{k+2}, $\ldots$, y_{T-k} and y_T). We use these to compute the autocorrelation (i.e., correlation between Y_t and Y_{t+k}) based on the expression in Equation (4.2) for both X and Z.

In terms of notation, we continue to use the same notation as in Hyndman and Athanasopoulos [20] as that reference could be used to dig deeper. For some $k \geq 0$, we estimate autocorrelation coefficient as

$$r_k = \frac{\sum_{t=k+1}^{T}(y_t - \bar{y})(y_{t-k} - \bar{y})}{\sum_{t=1}^{T}(y_t - \bar{y})^2} \tag{4.3}$$

which is a consequence of applying Equation (4.2) for X and Z to Y_t and Y_{t+k}. For the sake of clarification, notice in Equation (4.3) the numerator only uses the aforementioned $T-k$ pairs of time-series values separated by k. Also, the denominator can use all T values as simply the variance term is estimated.

The autocorrelation function (ACF) is a set of r_k values for various k with a maximum of a specified limit. Recall that most software use $T/2$ or $T/3$ as an allowable limit. For the one-hour data we go up to $k=40$ which corresponds to 5 days (of 8 hours sunlight). That corresponds to $k=160$ for 15-minute granular data. Figure 4.7 uses 1 month irradiance data from SolarTAC in Aurora, Colorado,

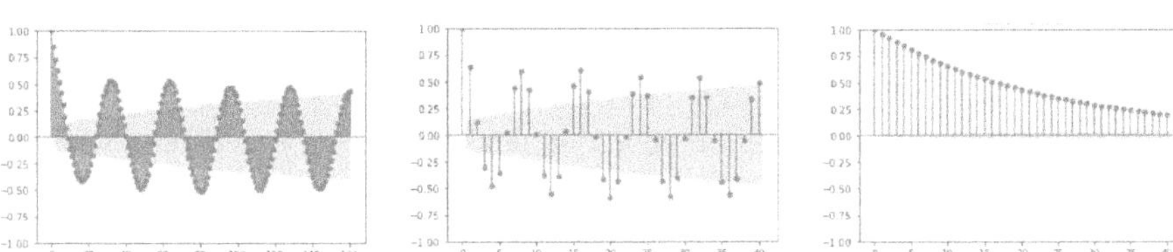

(a) ACF for 15-minute Avg. (b) ACF for 1-hour Avg. (c) ACF for Markov chain

Figure 4.7 Drop off in autocorrelation with lags for 1-month data

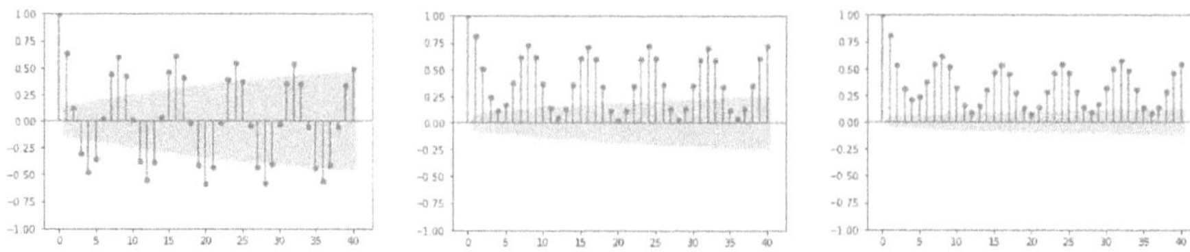

(a) ACF for 1 month data (b) ACF for 4 month data (c) ACF for 12 month data

Figure 4.8 Autocorrelation function for 1-hour average irradiance

United States (details are in Table 3.4) for the month of September 2016. The ACF for 15-minute interval is in Figure 4.7 (a) while that of 1-hour aggregate data is in Figure 4.7 (b).

Notice the sinusoidal nature of the data. From those two it is clear that whether it is 15-minute granularity or 1-hour, the maximum correlation is between now and the same time each of the next 4 days. The minimum is 4 data points later which in fact is negatively correlated. The light-blue sideways parabolic-looking shaded area is in fact the confidence interval. The way to interpret that is to state that the correlation (negative or positive) is significant beyond the confidence interval. In both the 15-minute granularity and the 1-hour granularity, there is a strong correlation that dies down to zero and then goes to negative and then rises back up. This periodic movement also indicates the seasonal periods. Although the cause of the sinusoidal is indeed due to the irradiance but the figure is not plotting that.

Now, as a comparison, we also consider simulation of a Markov chain. We simulated states of a Markov chain by letting it do one of three things in the next observation: increase by 1, decrease by 1, or remain the same with some given probabilities. The exception is when we reach the limits of 0 or 10 when we do not allow decrease or increase respectively. So the Markov chain stays between 0 and 10. We generated 100000 states and computed the autocorrelation function. That is depicted in Figure 4.7 (c). This has nothing to do with Figures 4.7 (a) or (b). The point is to show that in the Markov chain the correlations linger but also slowly disappear. Because we have so many data points, the confidence interval is just around zero. Notice that the Markov chain does not have a seasonal period.

Next consider Figure 4.8. Here we only use the hourly data. Note from all ACF plots when $k = 0$, the autocorrelation coefficient $r_0 = 1$. This can easily be derived by letting $k = 0$ in Equation (4.3). Figure 4.8 (a) is identical to Figure 4.7 (b) as both represent the ACF for 1-hour average irradiance data over 1 month. Now, in Figures 4.8 (b) and (c) we see what happens to the ACF if we use 4 months of data and 12 months of data. Notice the ACF plots are different. One reason is that there is a significant trend which is negligible in a 1-month sample. But it gets pronounced in 4 months and a lot worse in 12 months.

One thing to notice is that all the correlations in Figure 4.8 (b) and (c) are above zero, that means there is no negative correlation. This is not surprising because there is a lot more variability from month to month that within 5 days the correlation appears to be positive. Also, as we have a lot larger number of data points, the

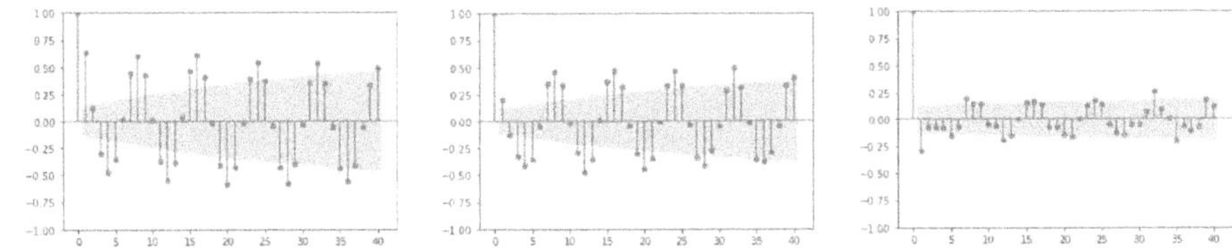

(a) ACF for raw data (b) ACF for first-order diff. (c) ACF for second-order diff.

Figure 4.9 Autocorrelation for 1-hour average irradiance over 1 month

confidence intervals are shrinking as we go from (a) to (c) in Figure 4.8. Also, notice how many points are outside the confidence interval. This tells us that there is a tremendous amount of non-stationarity due to seasonality and trend.

Recall from Section 3.2 that we could try and create stationary time-series by using first and second-order differencing shown in Figure 3.14. Now consider Figure 4.9. Here Figure 4.9 (a) corresponds to raw data which is identical to Figure 4.8 (a) which is identical to Figure 4.7 (b) as all three represent the ACF for 1-hour average irradiance data over 1 month. This is the raw data. Now we ask, what if we did a first-order differencing. Figure 4.9 (b) shows that first-order differencing has certainly brought several points to within the confidence interval, but there are still many outside. However, when we do a second-order differencing, the ACF in Figure Figure 4.9 (c) shows that most points are within and the data is nearly stationary. The only real dependence is when $k = 1$ and a few other ones like $k = 8$ which is the period, as expected and a few other points that are harder to explain. Also, even with second-order differencing there is the sinusoidal pattern which is expected due to the strong seasonality.

However, it is worthwhile asking if each lag is independently contributing to the forecast or is it the correlation an artefact of the fact that there is a relation between consecutive observations. The Markov chain example in Figure 4.7 (c) shows how much of an effect still persists over time. To address this issue, one typically uses another but related metric called partial-autocorrelation. The key idea is that we obtain the correlation between Y_t and Y_{t+k} (for some $k \geq 0$) but by controlling for the observations in between, i.e., $Y_{t+1}, \ldots, Y_{t+k-1}$. Mathematically, it is the conditional covariance over the conditional variances that we call the partial correlation. It is important to realize that for smaller lags (i.e., k) this is a fairly reasonable metric, but for larger k we need an excessively huge amount of data to span all possible combinations of control variables.

With that caveat, we proceed to explain the process of estimating the partial autocorrelation function (PACF). Similar to the ACF, the PACF for all integers k (such that $k \geq 2$) can be estimated from samples using

$$\phi_{k,k} = \frac{r_k - \sum_{i=1}^{k-1} \phi_{k-1,i} r_{k-i}}{1 - \sum_{i=1}^{k-1} \phi_{k-1,i} r_i} \tag{4.4}$$

where $\phi_{k,i} = \phi_{k-1,i} - \phi_{k,k}\phi_{k-1,k-i}$ for $1 \leq i \leq k-1$. The expression is computed recursively. For that, notice that $\phi_{0,0} = 1$ (since Y_t is perfectly correlated with Y_{t+k}

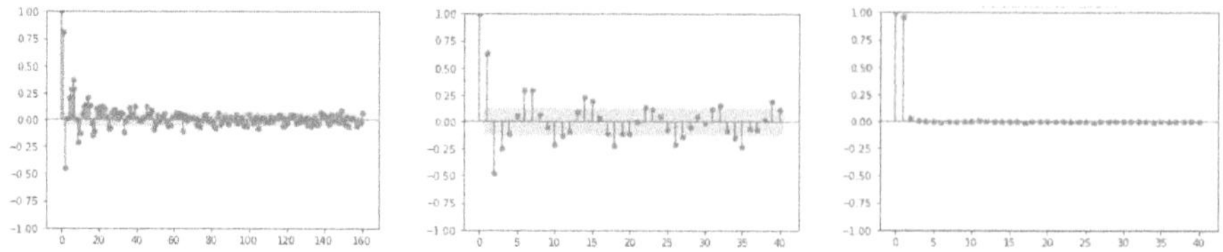

(a) PACF for 15-minute Avg. (b) PACF for 1-hour Avg. (c) PACF for Markov chain

Figure 4.10 Drop off in partial autocorrelation with lags

for $k = 0$). Then, $\phi_{1,1} = r_1$ because there is nothing between Y_t and Y_{t+1} and so the correlation is the same as partial correlation.

Now we consider Equation (4.4) for $k = 2$. By simply substituting, we have $\phi_{2,2} = (r_2 - \phi_{1,1}r_1)/(1 - \phi_{1,1}r_1)$. We can easily compute this as we have expressions for r_2, r_1, and $\phi_{1,1}$. Next we present for the case $k = 3$ in Equation (4.4). We get $\phi_{3,3} = (r_3 - \phi_{2,1}r_2 - \phi_{2,2}r_1)/(1 - \phi_{2,1}r_1 - \phi_{2,2}r_2)$. To compute $\phi_{3,3}$ we can compute r_3, we already have computed r_2, r_1, and $\phi_{2,2}$, but we need $\phi_{2,1}$. For that we use the expression below Equation (4.4), so $\phi_{2,1} = \phi_{1,1} - \phi_{2,2}\phi_{1,1}$. When we do $k = 4$, in a similar fashion we would need $\phi_{3,2}$ and $\phi_{3,1}$ that would need to be computed using terms we have already estimated. In this manner we build up for higher k.

Software packages with in-built functions are able to compute $\phi_{k,k}$ for any $k \geq 2$ using Equation (4.4). In Figure 4.10 (a) and (b) we present the PACF plots for SolarTAC in Aurora, Colorado, United States (details are in Table 3.4) for the month of September 2016. The Figure 4.10 (c) is the PACF for a 10-state Markov chain. In essence, Figure 4.10 is the PACF of the corresponding setting for ACF we saw in Figure 4.7. Unlike Figure 4.7 (a) and (b), in Figure 4.10 (a) and (b) there is a difference in the PACF between 15-minute average and 1-hour average, although the trend is similar. However, the 15-minute interval is too crowded to make any major insightful observations. However, from 4.10 (b), we can say that to forecast the hourly irradiance, we could use the previous three hours' irradiance as well as the previous sixth and seventh hour.

Hence, the major benefit with PACF is that we are able to determine which of the lag features to use in the forecasting models. In fact, in that sense it is more useful than the ACF because some of the significant lags in the ACF might not be useful. This is well depicted in the Markov chain results in 4.10 (c). Notice that the PACF is high for $k = 1$ and then goes to zero for $k > 1$. This is expected because of the Markov property which states that conditioned on the present, the future is independent of the past. So, all the PACF values when you condition on the intermediate states will be zero because of independence, the correlation would go off to zero. As expected for Markov chains, if we were to predict the next state, all we really need is the current state and nothing earlier.

An important aspect to note is that we are forecasting a day in advance. So we can only use features that lag 8 or higher in a 1-hour average time-series. So that is an aspect to think about while selecting features. In addition, there are still too many situations with sinusoidal patterns. This can be avoided by making the data

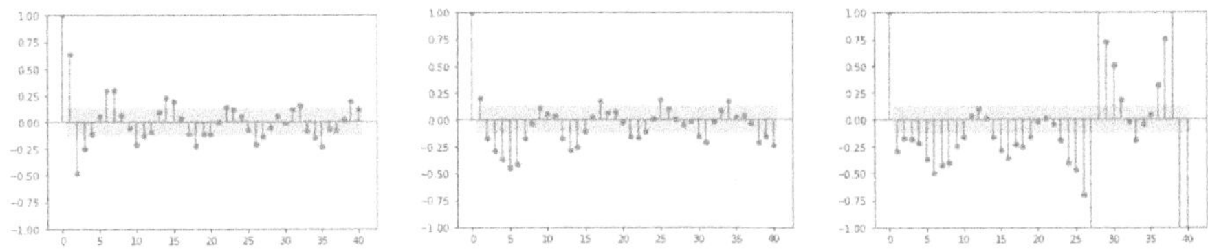

(a) PACF for raw data (b) PACF for First-order diff. (c) PACF for Second-order diff.

Figure 4.11 Partial autocorrelation for 1-hour average irradiance over 1 month

less seasonal. One way to do that is by differencing. As stated earlier in this section, details of differencing can be found in Section 3.2. Raw data, first-order differencing, and second-order differencing results are shown in Figure 4.11 (a), (b), and (c), respectively for PACF. The second-order differencing results in some instability or inaccuracies. One must be careful to check those before using the results. We wonder, is there another way for us to create stationary data (or nearly stationary)? We address that next.

To discuss the issue of stationary time-series as well as the ideal features to use in the methods to make forecasts 1-day in advance, we go back to the physical system and the forecasting approaches. The gradient boosting as well as the time-series approaches have to not only account for the uncertainty of cloud cover but also adjust for the time of day (seasonality) and the day of the year (trend). This makes forecasting quite challenging and we ask if we can reduce the variability in some way. Ideally, if we remove the seasonality and trend, thereby leaving a stationary process, we would be able to make more effective forecasts. One way to do that as we saw in the previous chapter is the concept of differencing (see Figure 3.14). When the underlying variability due to seasonality and trend cannot be explained, then differencing is an excellent option. However, here the seasonal variability and trend are mainly due to the earth's rotation and revolution. We explore that first and then study the issue of autocorrelations which form the foundations of time-series models.

The question we ask is: given the latitude and longitude of a location with a solar panel, on a cloudless day, can we forecast the irradiance at a given time and date? The answer is "yes" because this is after all the intensity of sunlight which can mathematically be represented. However, it involves several concepts from astronomy, trigonometry, and solar physics. The full treatment is beyond the scope of this book. Interested readers are suggested to look at some excellent resources including Duffie and Beckman [13] (solar geometry chapter), Smets et al [41] (solar radiation chapter), and Meeus [32] (equation of time chapter). We present a basic approach and take a data-driven approach to fine-tune the results.

Say that we are given the latitude (Ψ) and longitude (χ) of a location where we need to forecast the irradiance (without the effects of clouds) on a particular day of the year (*DOY*) at time τ. For example, suppose we wish to estimate the solar irradiance (not including cloud effects) in SolarTAC in Aurora, Colorado, United States (details are in Table 3.4) on September 20, 2017 as an average between 10am and 11am. From Table 3.4 we have $\Psi = 39.757$ and $\chi = -104.62$ with the negative

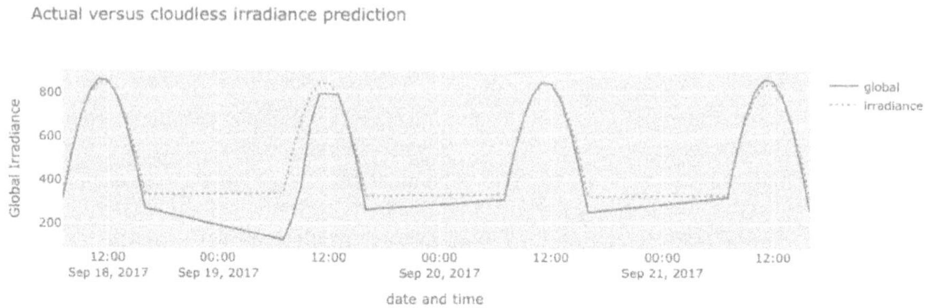

Figure 4.12 Observed irradiance ('global') versus clear sky predictions ('irradiance') over 4 reasonably-clear days

time to denote west (east would be positive). Also, $DOY = 263$ as it is 263rd day of a non-leap year. The time τ is computed as $10 + 0.5 + 0.0253$ because the average time is 0.5 hours past 10am, and that explains the 0.5 after the 10. But that time is at the US-Mountain Time based on the mean solar time corresponding to the longitude of 105 degrees West. So we need to adjust for 4 minutes per degree as SolarTAC is at $\chi = 104.62$ degrees and not at 105 degrees. For that, we need to add $(105 - 104.62) \times 4/60 \approx 0.0253$ hours to the time.

We call the time τ as the solar time at the location. Now using τ we obtain H, the solar time angle as $H = (\tau - 12) \times 15$ degrees where 12 is the 12th hour and 15 is the number of degrees per hour. We also compute the solar declination angle δ which is given as

$$\delta = 23.44 \cos\left(\frac{360}{365.24}(DOY - 173)\right)$$

where 360 corresponds to the number of degrees in a full circle, 365.24 the number of days in a year, and 173 the day of the year corresponding to the summer solstice. The number 23.44 in the above expression is the earth's tilt angle. Now, we can compute the irradiance (without accounting for cloud cover) as

$$1361K\left[\sin(\Psi)\sin(\delta) + \cos(\Psi)\cos(\delta)\cos(H)\right]$$

where 1361 is the solar constant.

In the above expression there is a constant K that needs to be explained. It is a parameter that we tune to adjust for the difference between the observed irradiance on a cloudless day and the predicted irradiance. This is necessary to account for other variables that were not considered in the calculations as well as the tilt of the solar panels. For the SolarTAC data $K = 0.8$ is used via a simple calibration made on cloudless days. For SolarTAC the actual versus clear-sky irradiance prediction is shown in Figure 4.12. Notice that the predictions are nearly perfect on most of the days. Cloud cover in early part of one of the days (September 19, 2017) caused the difference.

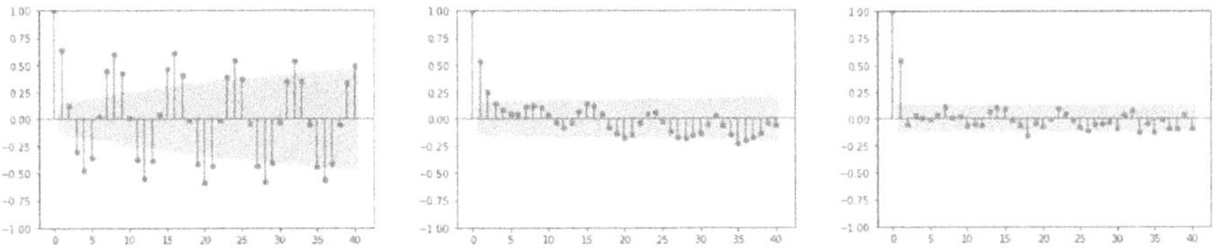

(a) ACF for raw data (b) ACF for scaled data (c) PACF for scaled data

Figure 4.13 Effect of Scaling on ACF and PACF for 1-hour average irradiance over 1 month

Also, there is a significant difference on the lower numbers because the data was collected for 10 hours accuracy is compromised due to data collection in the first and last hour. For our analysis we only use 8 hours of the day and hence this would not be an issue. Also notice that in two of the days the actual irradiance is higher than the cloudiness and that could be attributed to some diffused light that was reflected and hence added to the irradiance. The calculated irradiance does not take those effects into account. In summary, we can state that the calculated cloudless irradiance is capable of representing the seasonality and trend in the actual irradiance.

Remark

Since we have a way to estimate the cloudless irradiance at a location for a given time, we can scale the observed irradiance (corresponding to column `global` in the data sets) by dividing it by the cloudless irradiance. We create a new column in datasets and call it `scaled_global`. Thus we will use this time-series corresponding to the scaled data and predict what the scaled value would be. We would also obtain forecast accuracies for this scaled data only. It is relatively straightforward to multiply the prediction by the cloudless irradiance value at the location and time to get the forecasted irradiance. Before forging ahead with the autocorrelation of this time-series, it is worthwhile pointing out that the effect of clouds and diffused light is what causes variability.

In Figure 4.13 we present some results for scaled 1-hour average irradiance in SolarTAC with the goal of observing the effect of scaling data on ACF and PACF. Figure 4.13 (a) depicts the ACF for the raw (unscaled) time-series while Figure 4.13 (b) shows ACF for the scaled time-series. Notice from the two plots that in the scaled case after the first few lags, the remaining are inside the confidence interval, as compared to the unscaled data. However, there are still some sinusoidal patterns. This leads us to believe that while some huge seasonal patterns have been subdued, there still are some time of the day effects. This is mainly because the fit is not perfect at all times and certain times could be worse than others, and so the scaled values still cycle through the day.

Also notice from Figure 4.13 (c) that the PACF plot also has some wave patterns but largely within the confidence intervals. This is especially true for smaller lags. The important aspect to consider is that we actually need indicators beyond the first few lags because we predict 8 periods ahead. However, note that if we were

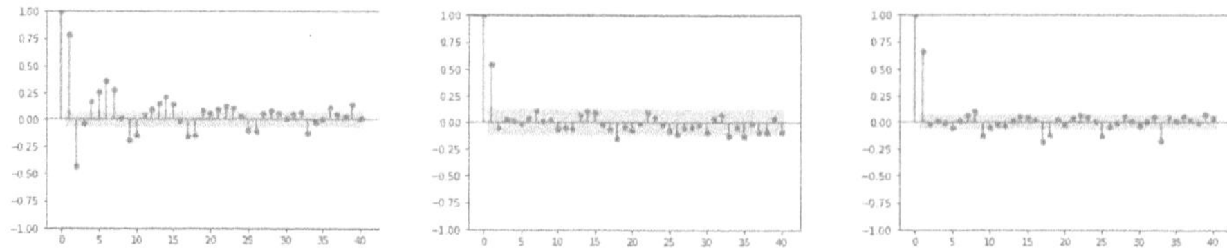

(a) PACF for raw data (3 mo.) (b) PACF for scaled data (1 mo.) (c) PACF for scaled data (3 mo.)

Figure 4.14 Effect of scaling on PACF for 1-hour average irradiance over 1 and 3 months

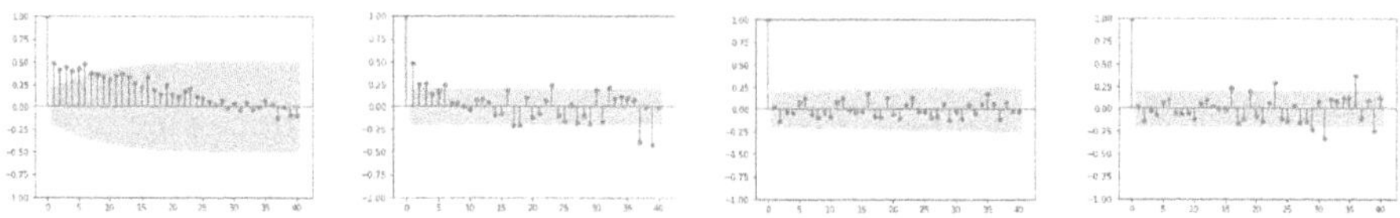

(a) ACF unscaled (b) PACF unscaled (c) ACF for scaled (d) PACF for scaled

Figure 4.15 Effect of scaling on ACF and PACF for data at 10am

to use this kind of a scaled data for the short-term forecasts in the previous chapter, the 1 hour ahead would be an excellent predictor. However, here although the scaling has certainly brought the variability to be low, it needs to be seen if the accuracy of forecasts continue to be good. Also, it may be that the exogenous variables such as weather (which is the main reason the difference between the cloudless and the cloudy conditions show a difference) would now play a more important role.

Next, let us compare just the PACF values between scaled and unscaled as well as between 1 month and 3 months. This study to understand the effect of scaling data on PACF is shown in Figure 4.14. Compare Figures 4.14 (a) and (b). Here we can see that the PACF is tremendously contained when we use scaled data. Further, the effect of time-of-the-day that causes seasonal variability is also greatly contained. Now, comparing Figures 4.14 (b) and (c) where the latter is over 3 months, notice that the confidence interval has shrunk due to the larger number of data points. On one hand we can see some good candidate points to use for the predictions such as 8 periods (i.e., 1 day) and 16 periods. On the other hand it may also be that scaling may have to be recalibrated. These aspects are important to consider while creating a forecasting tool.

As a final set of ACF and PACF plots, we consider a time series with 10am data alone. We simply filter out the 10am data as a representative sample time and see how the ACF and PACF would look like for both the scaled as well as unscaled data. That is shown in Figure 4.15. Notice from Figure 4.15 (a) that the correlation slowly dies down. It is crucial to note that the lag is for 40 days (and not 5 days like the earlier figures). Also, when we do not scale, the PACF for smaller lags do show much lesser dependence compared to using the whole time series. However, there is merit in considering the whole time series because consecutive hours of cloud cover can certainly be more useful in predicting irradiance than simply using 10am.

Further, Figure 4.15 (c) and (d) show ACF and PACF for scaled data and from 10am to 10am the next day, the data looks for all practical purposes stationary.

In summary, we have discussed exogenous variables and also ways to scale data to obtain stationary data as well as the right features. Next step is to describe ways to make the forecasts. The approaches using exponential smoothing and SARIMAX are identical to the previous chapter and hence are not repeated here. However, regression-trees and their ensembles (such as gradient boosting) were not explained in detail in the previous chapter. In the next section we provide some details. Then, in the final section we will provide comparison of various forecasting methods for the data sets we have. The reader can safely skip the next chapter (and revisit as needed) if the inner working of the regression tree approach is not of interest.

4.3 ENSEMBLE OF REGRESSION TREES

We have already introduced gradient boosting in the previous chapter (Section 3.3) and also visited it earlier in this chapter (Section 4.1). Gradient boosting is a type of ensemble of regression trees. However, we have not described any details of how it works and why it may be useful. In fact, different ensembles of regression trees are one of the most widely used methods for tabular data predictive analytics for their accuracy. It is often touted as the secret sauce in machine learning. When we can collect a vast amount of data and there is a need to predict several periods ahead, with the ability to use multiple exogenous features, the ensemble of regression trees outshine traditional methods of time-series forecasting.

The basis of gradient boosting (which is an ensemble of regression trees), naturally, is the topic of *regression tree*. So, in the next several paragraphs we describe regression-trees. Then we explain how we can use an ensemble of multiple regression trees. After that we show XGBoost which is gradient boosting on steroids. In the machine learning tools and software, it is common to call a regression tree as a type of *decision tree*. However, in the decision-analysis community the concept of decision-tree is different. For that reason we will simply call this regression tree. There is also a classification tree in machine learning which outputs a class (i.e., a categorical value) as opposed to a continuous value output by a regression tree.

To explain a regression tree, we start with a small data set with 2 features (so that it can be visualized easily on a graph). For that we consider three days of hourly solar irradiance data for Aurora CO (SolarTAC). This is the same location and same data set used earlier in this chapter. Our goal is to forecast the day-ahead scaled solar irradiance (details of scaling are in Section 4.2). We only use two features: Denver temperature forecast (a.k.a. Denv temp) in degrees Kelvin and the day-ahead hourly scaled irradiance data (1 seasonal period lag data, a.k.a. lags). The data is described in Table 4.1. The first column is simply an index to represent the row number. We will also not use the second column, which is the time-stamp for the date and hour.

The third column in Table 4.1 is the variable we wish to predict (so it is the y variable or response variable or dependent variable). The last two columns correspond to the two features we will use, i.e., Denver temperature and 1-day lag data. It is worthwhile noting that the scaled irradiance on 9/2/16 and 9/3/16 are the lags on 9/3/16

Table 4.1
Sample dataset to explain regression tree and ensembles

Index	datetime	scaled_global	Denv_temp	lags
a	9/2/16 8:00 AM	0.723820661	292.54	0.920982928
b	9/2/16 9:00 AM	0.6159811	291.2	0.955284483
c	9/2/16 10:00 AM	0.860241473	290.67	0.969001015
d	9/2/16 11:00 AM	0.826378452	289.95	0.978080696
e	9/2/16 12:00 PM	0.553728236	289.26	0.976360895
f	9/2/16 1:00 PM	0.845394979	289.66	0.972091013
g	9/2/16 2:00 PM	0.981795926	290.52	0.91204926
h	9/2/16 3:00 PM	0.938795972	290.78	0.945252678
i	9/3/16 8:00 AM	0.920494171	288.75	0.723820661
j	9/3/16 9:00 AM	1.020041799	288.4	0.6159811
k	9/3/16 10:00 AM	0.933964695	287.85	0.860241473
l	9/3/16 11:00 AM	0.999310499	288.36	0.826378452
m	9/3/16 12:00 PM	1.054452256	287.73	0.553728236
n	9/3/16 1:00 PM	0.946686228	288.22	0.845394979
o	9/3/16 2:00 PM	0.653237863	289.72	0.981795926
p	9/3/16 3:00 PM	0.790649024	293.13	0.938795972
q	9/4/16 8:00 AM	0.951352593	287.68	0.920494171
r	9/4/16 9:00 AM	0.849309704	287.03	1.020041799
s	9/4/16 10:00 AM	0.481302952	285.6	0.933964695
t	9/4/16 11:00 AM	0.662558082	285.39	0.999310499
u	9/4/16 12:00 PM	1.061847508	285.61	1.054452256
v	9/4/16 1:00 PM	0.519071854	286.43	0.946686228
w	9/4/16 2:00 PM	0.593879813	290.6	0.653237863
x	9/4/16 3:00 PM	0.946359426	293.03	0.790649024

and 9/4/16, respectively. This is what we mean by 1-day lag. For example, on 9/4/16 for 8am (index q), the features used to predict the scaled irradiance will be the forecasted Denver temperature (which is 287.68 Kelvin) and 1-day lag scaled irradiance is 0.920494171 (which is the scaled irradiance on 9/3/16 for 8am, i.e., index i).

As a start, we first plot the 24 lines of data in Table 4.1 (corresponding to three days of hourly data across sunlight hours of 8am to 4pm) in Figure 4.16. The X-axis in Figure 4.16 corresponds to the column `Denv_temp` in Table 4.1. The Y-axis in Figure 4.16 corresponds to the column `lags` in Table 4.1. The 24 data points for the two features are plotted using a scatter plot. The text on top of each point in the scatter plot is the scaled global irradiance rounded off to the first decimal. For the moment it is not necessary to exactly map each scatter plot point to a line on Table 4.1, but we will do so subsequently.

The key idea is to see if we could somehow form sets of points grouped together so that when a future prediction has features close to a group of points, we can predict the future irradiance as a function of the irradiance of the points in the group. This in some ways is similar to a regression where we have a mathematical relation

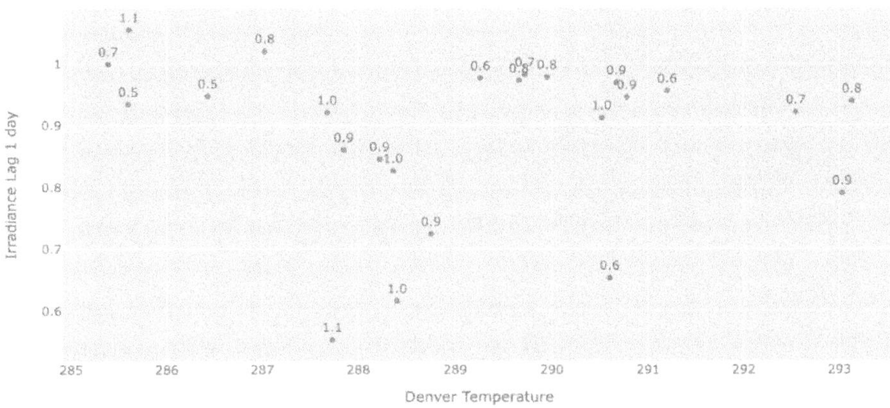

Figure 4.16 Scaled Global Irradiance (rounded to 1 decimal) at various temperature and lag points

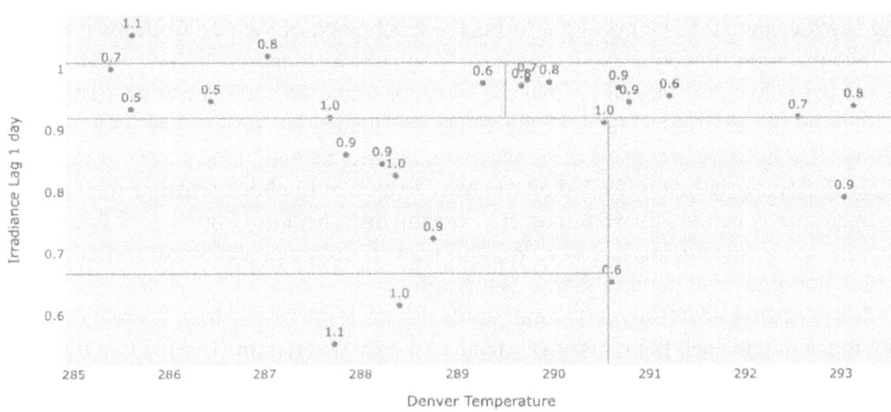

Figure 4.17 Grouping Scaled Global Irradiance (rounded to 1 decimal) into Regions

between the feature values and the predicted variable. In other words, given a feature we have the ability to predict what values the dependent variable would take. But, the approach is very different from a general linear regression with non-linear features. However, the approach for both is statistical without much regard for the physical relationship between the variables.

To form groups of points so that each group forms a region, we use a tree-based method. For that reason this is called a regression *tree*. Figure 4.17 describes a way to group the points we saw in Figure 4.16 into 6 regions. These 6 rectangular regions can be derived using a tree structure. The way this tree works is the root checks whether the 1-day lag (y-axis) is less than 0.92 or not. If the 1-day lag is smaller than 0.92, then the next branch of the tree is to check if the Denver temperature (x-axis)

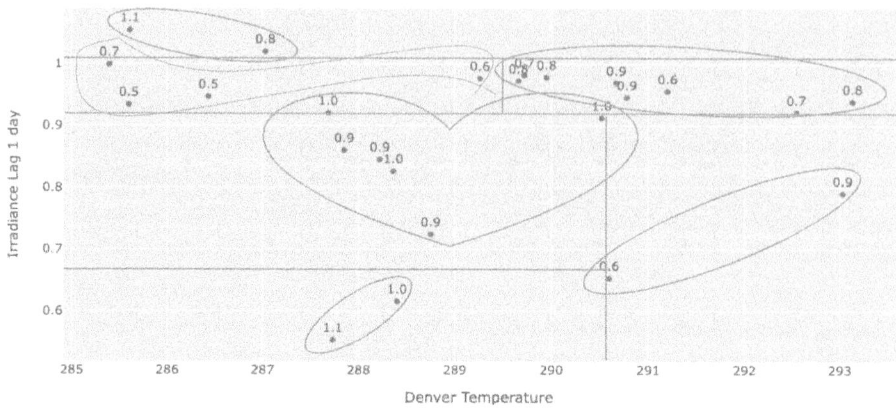

Figure 4.18 Mapping points to regions to show where boundary points are mapped

is below 290.56 or not. If it is below, then in the next branch we check if the 1-day lag is lesser than 0.67 or not. This forms the left branch of the tree from the root.

For the right branch at the root, all the points have 1-day lag values greater than 0.92 (the values are picked so that the issue of equal to does not arise). The next branch of the tree is to check if the 1-day lag values are lesser than 1.01 or not. Notice that here we are again using the same feature we had in the parent of the tree. If the 1-day lag is below 1.01 then the next branch is to check whether the Denver temperature is below 289.46 or not. It is worthwhile checking Figure 4.17 how the 6 regions are formed using this tree. We will later discuss a procedure to form the tree and at that time also depict the tree as a graph.

It is worthwhile noting that some points appear to be on the line, however that is not the case and each point belongs to one and only one region. To illustrate this, the points are grouped using shapes in Figure 4.18. So, among the six regions there are three with only two points apiece. Although the divisions may not make sense from a physical standpoint, notice that the points in a region have somewhat similar scaled global irradiance values (shown to 1-decimal accuracy). This allows us to guess that future values in each region will take similar values as those already in a region.

In regression trees, the mapping between the values of the points in a region to a predicted value is simply the average of all the training points in that region. Figure 4.19 shows $\hat{y}$ as the estimated values in each region. It is important to note that all the calculations are done without rounding off, but what is presented in Figure 4.19 are rounded to 1 or 2 decimals. For example, the lowermost and left region has two points 1.0 and 1.1 whose true values are 1.020041799 and 1.054452256, respectively. By averaging those two values, we get 1.0372470275 which is rounded off to two decimal places as $\hat{y} = 1.04$ as shown in Figure 4.19. In a similar manner, the estimates in other regions are obtained by averaging over points in those regions.

Thus, based on Figure 4.19, if we used it as a regression tree to make predictions, then a forecasted Denver temperature of 288 Kelvin and scaled irradiance a day

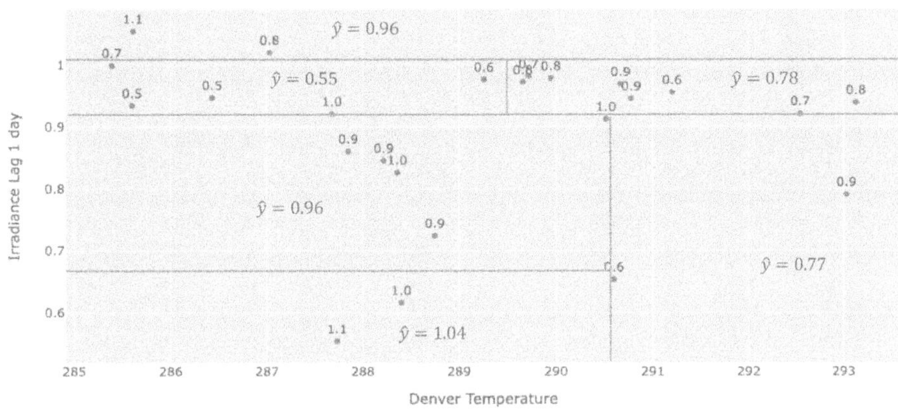

Figure 4.19 Estimated scaled irradiance in each region

ago in that hour of 0.6 would yield a prediction of $\hat{y} = 1.04$ as scaled irradiance. Likewise, if the forecasted Denver temperature is 292 Kelvin and the 1-day lag scaled irradiance is 0.8, the predicted scaled irradiance would be $\hat{y} = 0.77$ for that hour. However, for the same 1-day lag scaled irradiance of 0.8, if the forecasted Denver temperature was 287, then $\hat{y} = 0.96$ would be the predicted scaled irradiance. Also, for a 1-day lag irradiance of 1.0, the predicted scaled irradiance $\hat{y}$ value would be 0.55 and 0.78 for a forecasted Denver temperature of 286 and 291, respectively. Finally, if the 1-day lag scaled irradiance is 1.05, then irrespective of the Denver temperature, the scaled irradiance will be forecasted as $\hat{y} = 0.96$.

Having described how to use a regression tree to make a prediction, next we see how this tree was constructed. For that, we redraw Figure 4.17 by replacing the scaled irradiance (rounded off to 1 decimal) by the index value (see Table 4.1), and this way we could refer to a data point that uniquely refers to a line in the data set shown in Table 4.1. This redrawn graph is shown in Figure 4.20. We first show how these points are mapped to a tree and then we will see how the tree was constructed. However, all the discussions about predictions are the same as before, this Figure 4.20 is just replacing the target value by the index with the aim to effectively illustrate.

In Figure 4.21 we present the scatter graph with six regions shown in Figure 4.20 as a binary tree with six leaves (the oval shaped objects with indices a to x corresponding to the index values in Table 4.1). We also show the predictions $\hat{y}$ corresponding to the six leaves. The topmost rectangular object (with `lags` < 0.92 written on it) is the root of the tree. The other four rectangular objects can be considered as branches. This is a binary tree as the root and branches yield a binary result of True or False. In this manner it would be possible for one to quickly go to the leaf corresponding to the features presented to make a prediction.

The height of the tree in Figure 4.21 is 3. This is also called *depth*. The reason the height is calculated as 3 is because it is the maximum number of hops between the root and leaf, i.e., the maximum number of True/False checks to be performed before

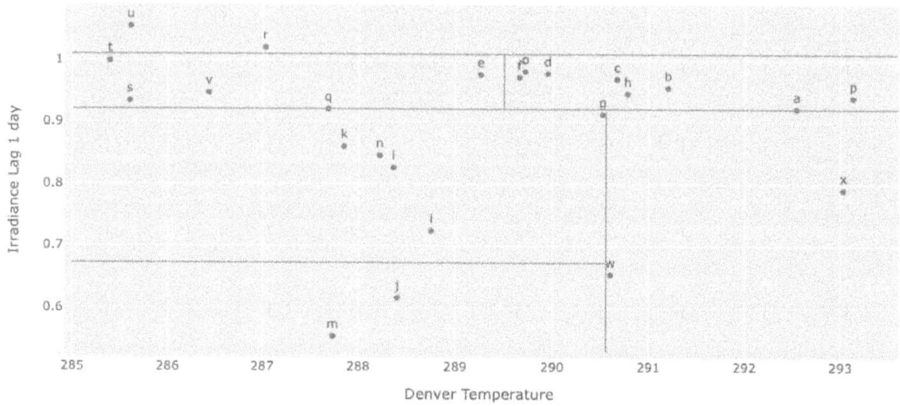

Figure 4.20 Index values of data set mapped to regions

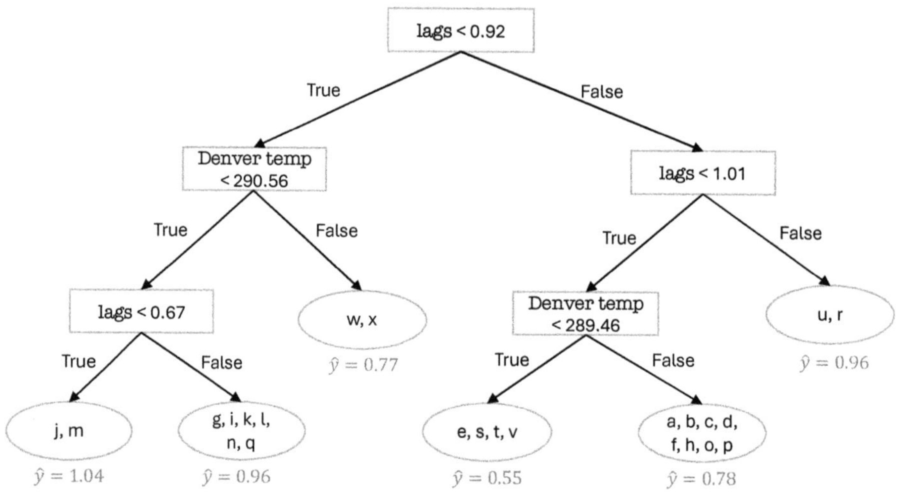

Figure 4.21 Regression Tree Depicted as a Binary Tree

reaching a leaf. Also notice that each leaf has at least 2 samples. These are some of the restrictions (there are many more) that can be stated in python libraries. In fact, upon using the library for decision-tree-regressor in `scikit-learn.org` libraries for the dataset in Table 4.1 stating `max_depth` as 3 and `min_samples_leaf` as 2, we obtain the tree in Figure 4.21.

Now we explain the process of obtaining the tree in Figure 4.21. We build a tree by placing the restrictions that the tree cannot have a depth of greater than 3 (so `max_depth` is set as 3) and each leaf must have at least 2 samples (so `min_samples_leaf` is set as 2). With this in mind, we ask what should be the root? Should we say `lags < L` or `Denver temp < T`? And what should L or T be? The answer is we try all values of L and T (there actually are only 21 possible values for each, we will see soon). Among all the 42 options, the one that has the smallest sum of squared error (SSE) will be the choice.

To explain SSE, we first say that when we consider a root, all the points y that will result in `True` will be predicted as $\hat{y}$ which is the mean of the points in the True branch of the tree. Likewise all the remaining points will be predicted as the average of the False branch of the tree. Then the SSE is simply the sum of squares of the y value minus the $\hat{y}$ value of that point (which is the error in estimation for that point), summed over all the points. To illustrate this, let us consider the root of the tree in Figure 4.21.

Pretend that the only thing we have on the tree is the root with `lags < 0.92` and nothing else. Clearly, the points g, i, j, k, l, m, n, q, w, x will satisfy as "True" and the remaining points a, b, c, d, e, f, h, o, p, r, s, t, u, v will fall to the right of the root under "False". Here we only have two leaves (note from Figure 4.21, these form the eventual leaves on either sides). The average irradiance across the 10 points g, i, j, k, l, m, n, q, w, x is 0.9348 (so the irradiance forecast for all those points is $\hat{y} = 0.9348$).

Likewise, the average irradiance across the 14 points a, b, c, d, e, f, h, o, p, r, s, t, u, v is 0.7416 (so the irradiance forecast for all those points is $\hat{y} = 0.7416$). Now to get the SSE it is equal to

$$(y_a - \hat{y}_a)^2 + (y_b - \hat{y}_b)^2 + \ldots + (y_x - \hat{y}_x)^2$$

where $y_a, y_b, \ldots, y_x$ is the scaled global irradiance corresponding to points a, b, ..., x respectively and $\hat{y}_a$, $\hat{y}_b$, ..., $\hat{y}_x$ are the estimated global irradiance for those same points. Using a rounding to 4 decimals we can compute SSE as $(0.7238 - 0.7416)^2 + (0.6160 - 0.7416)^2 + \ldots + (0.9466 - 0.9348)^2 = 0.5151$ by noticing that we used $\hat{y}$ depending on whether the root condition is satisfied or not.

If we were to look at Table 4.1, we will notice that point q has a `lags` value greater that 0.92 and yet it is in the "True" side. That is because, we had rounded off the threshold L in `lags < L` to 2 decimals. Using 5 decimals we would actually code the root as `lags < 0.92074` and that would result in point q coming as False but point a as True. In fact the 0.92074 is taken as the average of the points for a and q which are 0.92098 and 0.92049 respectively. So, these threshold values are determined as the midpoint between successive values in a sorted list of `lags` values.

We first sort the 24 data points according to `lags` which gives us the order (from smallest `lags` to largest) m, j, w, i, x, l, n, k, g, q, a, s, p, h, v, b, c, f, e, d, o, t, r, u. So the options are to create a threshold between j and w, w and i, i and x, and so on, until t and r. The reason we do not create between the first two or the last two is that we need at least 2 points in a leaf. There are indeed the 21 ways we can split the lags into two groups. We compute the SSE for each of the 21 points. This is depicted in Table 4.2.

From Table 4.2, the best place to insert a threshold is between a and q, which we saw earlier. However, we need to also check to see if the SSE is lower for the threshold on Denver temperature to determine what should be the root (if we choose lags, then it is indeed `lags < 0.92074`). Next we try the 21 possible values of T

Table 4.2
Options for splitting the "lags" column at root of a regression tree

Split between	j & w	w & i	i & x	x & l	l & n	n & k	k & g
$\hat{y}_{left}$	1.037	0.889	0.897	0.907	0.922	0.926	0.927
$\hat{y}_{right}$	0.803	0.812	0.807	0.800	0.789	0.779	0.770
SSE	0.632	0.717	0.706	0.687	0.652	0.627	0.601
Split between	g & q	**q & a**	a & s	s & p	p & h	h & v	v & b
$\hat{y}_{left}$	0.933	**0.935**	0.916	0.879	0.873	0.877	0.853
$\hat{y}_{right}$	0.756	**0.742**	0.743	0.765	0.762	0.745	0.770
SSE	0.556	**0.515**	0.555	0.654	0.661	0.630	0.694
Split between	b & c	c & f	f & e	e & d	d & o	o & t	t & r
$\hat{y}_{left}$	0.839	0.840	0.840	0.825	0.825	0.817	0.810
$\hat{y}_{right}$	0.789	0.779	0.768	0.811	0.807	0.858	0.956
SSE	0.720	0.715	0.709	0.732	0.732	0.729	0.694

Table 4.3
Options for splitting the "Denver temp" column at root of a regression tree

Split between	s & u	u & v	v & r	r & q	q & m	m & k	k & n
$\hat{y}_{left}$	0.572	0.735	0.681	0.715	0.754	0.797	0.814
$\hat{y}_{right}$	0.845	0.835	0.850	0.850	0.845	0.832	0.826
SSE	0.596	0.707	0.638	0.660	0.696	0.727	0.732
Split between	n & l	l & j	j & i	i & e	e & f	f & o	o & d
$\hat{y}_{left}$	0.829	0.846	0.862	0.867	0.843	0.843	0.830
$\hat{y}_{right}$	0.818	0.805	0.789	0.778	0.798	0.793	0.809
SSE	0.732	0.723	0.701	0.685	0.721	0.719	0.730
Split between	d & g	g & w	w & c	c & h	h & b	b & a	a & x
$\hat{y}_{left}$	0.830	0.839	0.825	0.827	0.833	0.822	0.818
$\hat{y}_{right}$	0.806	0.781	0.813	0.803	0.769	0.820	0.869
SSE	0.730	0.717	0.732	0.731	0.720	0.733	0.728

to see the best option for `Denver temp` < T. For that we sort the 24 data points according to `Denver temp` from smallest to largest temperature values. That results in the order t, s, u, v, r, q, m, k, n, l, j, i, e, f, o, d, g, w, c, h, b, a, x, p. Just like for the lags, here too we have 21 places where we can split into two groups by cutting between s and u, u and v, and so on until a and x. Table 4.3 summarizes the results.

The smallest SSE for this is 0.572 which is higher than what we saw in Table 4.2 for the 1-day lag feature. But before we decide to use `lags < 0.92074` as the root of the regression tree, we need to ask if it is justified (a step we will undertake while creating the branches as well). If we were to simply use all the 24 data points and that their average of 0.822 is what we will use as $\hat{y}$ for all the points, then the SSE becomes 0.733 which translates to an MSE (mean-squared error) of 0.0305 which is called the *impurity* of the model.

Now, by creating a tree with child node corresponding to whether or not `lags < 0.92074`, the SSE becomes 0.515 which is smaller than 0.733, and in fact it is the split that results in the *maximum* information gain. We formalize that next for any parent node p of a decision tree with left child node l and right child node r so that there are N_p, N_l, and N_r data points in the parent node, left child node, and right child node respectively, with $N_l + N_r = N_p$. One typically uses MSE for impurity and let I_p, I_l, and I_r be the impurity corresponding to the parent, left, and right child nodes. Then, the information gain is given by

$$I_p - \frac{N_l}{N_p} I_l - \frac{N_r}{N_p} I_r$$

which is essentially the difference in SSE before and after the split (as SSE is MSE times the number of data points).

Clearly, the split `lags < 0.92074` results in the maximum information gain. So we make that split and create a tree with two child nodes. Now, this tree has a depth of 1 and also satisfies the condition that the leaf nodes must have at least 2 samples. Next we ask, can we split the child nodes? The answer to that is yes, provided it results in an information gain and has at least 2 samples in the leaves (the depth would be a maximum of 2). Of course, we wish to greedily select the split that results in maximum information gain.

Consider the left child node of the split `lags < 0.92074`. It has the points g, i, j, k, l, m, n, q, w, x (since for these `lags < 0.92` will be "True"). We go back to the two features 'lags' and 'Denver temp' (in the case where there are more features, we would have to consider all of them). Among all the 7 thresholds for the 10 points across both features (note that we have already sorted in the previous split) we find the split that results in the maximum information gain. In this case, that ends up being `Denver temp < 290.56` with an SSE of 0.0771 which is a significant information gain with w and x forming the right child and the remaining the left child.

Since this right child with points w and x has reached the minimum samples of 2, it will not be split any more and remain a leaf. The left child node note has the 8 points g, i, j, k, l, m, n, q which we can split by considering both lags and Denver temp. Among all the feasible splits, `lags < 0.67` results in the maximum information gain that corresponds to an SSE of 0.00499 (splitting into left child points j and m, and the remaining with the right child). Now, this cannot be split any further as we have reached the maximum depth of 2. It is worthwhile seeing how this pans out in the left part of Figure 4.21.

Now we briefly explain the right part of Figure 4.21. Recall the 14 points that resulted in False in the root node `lags < 0.92074` which are considered next. Trying all possible feasible splits across both lags and Denver temp., we get the smallest SSE as 0.2634 corresponding to `lags < 1.01` for which the right child node has two points u and r. The left child node is further split using `Denver temp < 289.46` resulting in an SSE of 0.1026 which corresponds to the largest information gain. It splits it into a left child node corresponding to the points e, s, t, and v while the remaining points are with the right child node. Thus the tree in Figure 4.21 is constructed.

RANDOM FOREST AND GRADIENT BOOSTING

It is possible to extend the concept of a regression with 2 features (such as the one in Figure 4.21) with one with multiple features. Essentially we perform a split across a feature that results in maximum information gain. We continue until either a child node cannot be split any more (as the minimum number of samples in leaf is reached) or the maximum depth is reached. If neither are specified, it would result in a regression tree that suffers from over-fitting as it fits the training data perfectly but is likely to perform poorly with testing data. So we wish to avoid this. In addition, notice that the regression tree is built greedily taking advantage of the maximum information gain at each stage.

Thus a single regression tree is unlikely to perform strongly and the key idea is to use a collection of trees or ensemble of trees as it is called. Although we would not use *random forest* much in this book, its ease of explanation makes is a good first step to consider in building an ensemble. Say that our training data has n samples and m features to predict the target variable. Using these, instead of building a single regression tree, the idea in ensemble of trees is to build a set of B trees. For random forest, a random subset of the m features are selected (usually $\sqrt{m}$) and a regression tree is built. We repeat the process B times and B trees are built, and together called a random forest.

The key idea is that each of the B trees would make a prediction, and the output would be an average of the B predictions. This is an effective way and works particularly well when m is large, which in our situation is not the case (hence we do not use the method). However, it is worthwhile to point out that we can also sample from a subset of the n rows of data to build the tree. Thus the trees built are different and the issue of greedy selection is averaged out typically. Thus random forest is an effective way to build an ensemble of B trees. Usually B is about 50 to 100. In some sense the trees in the random forest can make predictions in parallel.

Gradient boosting is another ensemble method where B trees are built serially as opposed to parallel in *random forest*. Since we have seen a code for *gradient boosting*, we start there. In Figure 4.5, consider Cell 9. Line 5 in that cell essentially states that we use the gradient boosting regression and build $B = 100$ trees (`n_estimators` is the same as B) in series. Each tree is a regression tree with a `max_depth` of 5 and at least 2 samples in each leaf node. Then in line 7 we ask it to fit a gradient boosting ensemble of trees. Subsequently in Cell 10 of Figure 4.5 we show how the model is tested using predictions made in line 8 of Cell 10.

The key idea of the sequence of trees is the use of successive residuals. The first tree is the standard regression tree we saw earlier. For the ith sample say that y_i is the target value while $\hat{y}_i$ is the predicted value (depending on which region i falls in). Then the residual (measuring the error) is $y_i - \hat{y}_i$ for sample i. The next tree is built with these residuals, i.e., we replace y_i by $y_i - \hat{y}_i$. Then we build another tree and make predictions. Then compute the residual for each sample for that tree. This process continues for all the B trees. For the testing data, the predictions are added. More details are not provided as it is just a special case of XGBoost that we will describe next.

XGBOOST

XGBoost is an ensemble binary-tree based technique and stands for eXtreme Gradient Boosting. It can be used for both regression and classification, although in this book we only use regression. Hence that is described briefly. The scheme was originally developed by Tianqi Chen (see Chen and Guestrin [10]). It is based on a decision-tree framework and is non-parametric. A vast majority of Kaggle competitions' winning entries use XGBoost. Only few details are provided in this book and the reader is encouraged to watch Tianqi Chen's talk [9] and Trevor Hastie's talk [15] on this topic. There are also many tutorials and introductory write ups available online.

To explain how XGBoost works, we consider a generic tabular data set with Y as the dependent variable and $X_1, X_2, \ldots, X_m$ as m independent variables. Notice that for the solar forecasting problem, Y is indeed a column of scaled irradiance observed in an hourly fashion and X_j for $j = 1$ to m are various features such as day ahead lag, day ahead plus one hour lag, 2-day ahead lag, forecasted temperature, forecasted cloud cover, etc. For the purposes of explaining the approach we keep it generic. Earlier in this section we showed how to create a regression tree, given a dependent variable and a set of features. Here we focus on building a sequence of B regression trees.

Recall from the regression tree explanation that each row in the tabular data gets assigned to a region of the regression tree, and the regions are built in a greedy fashion based on the m features. So, generically we write down the forecast of the residuals b^{th} regression tree for the ith data point as a function $f_b(\mathbf{x}_i)$ for all $b \in \{1,2,\ldots,B\}$ and $i = 1$ to n, where $\mathbf{x}_i$ is an m-dimensional vector corresponding to the ith row of all the m features. In the random forest and gradient boosting that we presented earlier, $f_b(\mathbf{x}_i)$ was simply the average of all the samples that were with i in the leaf of the tree. For example, in Figure 4.21, if i corresponds to point q and $B = 1$, then $f_b(\mathbf{x}_i)$ for the only tree is 0.96, as seen in that figure.

However, in XGBoost, $f_b(\mathbf{x}_i)$ is not the average of all the points on the leaf of the tree, but it is a weight w_l that is computed as a common value for all the points i on the leaf l. That part is the same but it is not necessary that w_l is the average value of all the points on leaf l. Next we explain how $w_l = f_b(\mathbf{x}_i)$ is computed, given a tree b and a mapping of i to l. Typically, the prediction is made for point i as

$$\hat{y}_i = \overline{y} + \sum_{b=1}^{B} f_b(\mathbf{x}_i)$$

for all $i = 1$ to n, and $\overline{y}$ is the mean of all the y_i values such that the first tree is built simply using the residuals with the zeroth prediction as $\overline{y}$. Since $f_b(\mathbf{x}_i)$ is the predicted residual for point i, the prediction for point i in the bth tree is the $\hat{y}_i^{(b)}$ and it is computed due to residuals as $\hat{y}_i^{(b-1)} + f_b(\mathbf{x}_i)$. Although $w_l = f_b(\mathbf{x}_i)$, we need to be careful to use the $f_b(\mathbf{x}_i)$ when we refer to point i and w_l when we refer to leaf l.

For the moment until we find the best tree, we assume that we have a mapping from i to l for all i from 1 to n and l from 1 to T_b for a given binary tree b. Subsequently, we will obtain the best tree b. To compute w_l, i.e., $f_b(\mathbf{x}_i)$ under the mapping in the tree structure from i to l, we consider the loss function corresponding to the bth tree as

$$\mathscr{L}_b = \sum_{i=1}^{n} \ell(y_i, \hat{y}_i^{(b)}) + 0.5\lambda \sum_{l=1}^{T_b} w_l^2 + \gamma T_b \tag{4.5}$$

where λ and γ are constants used for regularization, T_b is the number of leaf nodes in the bth tree, and the error function $\ell(y_i, \hat{y}_i^{(b)})$ describes how we wish to measure the error $y_i - \hat{y}_i^{(b)}$. The XGBoost method allows for any error function but the most common one is

$$\ell(y_i, \hat{y}_i^{(b)}) = (y_i - \hat{y}_i^{(b)})^2$$

which when added across i becomes the usual sum of squared errors. We wish to find the residual $f_b(\mathbf{x}_i)$ that minimizes the loss function $\mathscr{L}_b$.

We substitute $\hat{y}_i^{(b)}$ with $\hat{y}_i^{(b-1)} + f_b(\mathbf{x}_i)$ in $\mathscr{L}_b$, noticing that $\hat{y}_i^{(b-1)}$ has already been optimally selected in the previous tree $b-1$. So given the tree b we now obtain the residuals $f_b(\mathbf{x}_i)$ (i.e., w_l) which is $\hat{y}_i^{(b)} - \hat{y}_i^{(b-1)}$. We use a Taylor series expansion for $\ell(y_i, \hat{y}_i^{(b)})$ as

$$\ell(y_i, \hat{y}_i^{(b)}) = \ell(y_i, \hat{y}_i^{(b-1)} + f_b(\mathbf{x}_i)) \approx \ell(y_i, \hat{y}_i^{(b-1)}) + g_i f_b(\mathbf{x}_i) + \frac{1}{2} h_i f_b^2(\mathbf{x}_i)$$

where g_i is the first derivative of $\ell(y_i, y)$ with respect to y taken at $y = \hat{y}_i^{(b-1)}$ and h_i is the second derivative of $\ell(y_i, y)$ with respect to y taken at $y = \hat{y}_i^{(b-1)}$. The approximation is *exact* when $\ell(\cdot,\cdot)$ is the squared error $\ell(y_i, \hat{y}_i^{(b-1)} + f_b(\mathbf{x}_i)) = (y_i - \hat{y}_i^{(b-1)} - f_b(\mathbf{x}_i))^2$ for which $g_i = -2(y_i - \hat{y}_i^{(b-1)} - f_b(\mathbf{x}_i))$ and $h_i = 2$.

For the remainder of this section, we let $\ell(\cdot,\cdot)$ be a generic convex function. Using the expressions in the previous paragraph, we can rewrite the loss function for tree b, $\mathscr{L}_b$, in Equation (4.5) as

$$\mathscr{L}_b = \sum_{i=1}^{n} \left(\ell(y_i, \hat{y}_i^{(b-1)}) + g_i f_b(\mathbf{x}_i) + \frac{1}{2} h_i f_b^2(\mathbf{x}_i) \right) + 0.5\lambda \sum_{l=1}^{T_b} w_l^2 + \gamma T_b$$

which we wish to minimize. Dropping the constant term $\ell(y_i, \hat{y}_i^{(b-1)})$, we rewrite the loss function as

$$\overline{\mathscr{L}_b} = \sum_{i=1}^{n} \left(g_i f_b(\mathbf{x}_i) + \frac{1}{2} h_i f_b^2(\mathbf{x}_i) \right) + 0.5\lambda \sum_{l=1}^{T_b} w_l^2 + \gamma T_b.$$

Now, letting L_l denote the set of samples that fall under leaf l in tree b so that for all $i \in L_l$ the estimate of the residuals is $f_b(\mathbf{x}_i) = w_l$, we can express the above loss function as

$$\overline{\mathscr{L}_b} = \sum_{l=1}^{T_b} \left((\sum_{i \in L_l} g_i) w_l + \frac{1}{2} (\sum_{i \in L_l} h_i + \lambda) w_l^2 \right) + \gamma T_b \tag{4.6}$$

which we need to minimize to obtain w_l for all $l \in \{1, 2, \ldots, T_b\}$.

For a given tree structure where the mapping from samples to leaves are known (i.e., the sets L_l for all $l \in \{1, 2, \ldots, T_b\}$ are given), we can simply use the gradient of the loss function to obtain the optimal weights as

$$w_l^* = -\frac{\sum_{i \in L_l} g_i}{\sum_{i \in L_l} h_i + \lambda}$$

for all $l \in \{1, 2, \ldots, T_b\}$. Interestingly, when $\ell(\cdot, \cdot)$ is the squared error $\ell(y_i, \hat{y}_i^{(b-1)} + f_b(\mathbf{x}_i)) = (y_i - \hat{y}_i^{(b-1)} - f_b(\mathbf{x}_i))^2$, and $\lambda = 0$, then w_l^* is the average of all the points in L_l (this should not be surprising as it is well known that the average of a set of numbers minimizes the sum of squared errors when used to estimate those numbers). Thus when $\lambda > 0$ or $\ell(\cdot, \cdot)$ is a function different from the squared error, we can use w_l^* as shown above.

Replacing w_l by the expression for w_l^* in Equation (4.6), the minimum value of the loss function is

$$\overline{\mathscr{L}_b^*} = -\frac{1}{2} \sum_{l=1}^{T_b} \frac{\left(\sum_{i \in L_l} g_i\right)^2}{\sum_{i \in L_l} h_i + \lambda} + \gamma T_b \tag{4.7}$$

for a given tree structure that maps all the samples i from 1 to n to one of the leaves L_1, L_2, ..., L_{T_b}. However, we also want to obtain the binary regression tree which among all trees minimizes the loss function in Equation (4.7). This process is identical to what was done earlier to obtain the binary regression tree in Figure 4.21. However, there are some differences. To create the bth tree in XGBoost, instead of used MSE (which was used to obtain the tree in Figure 4.21), here we use the optimum loss function in Equation (4.7). Also, the target variable for sample i in the bth tree is the residual $y_i - \hat{y}_i^{(b-1)}$.

Since the process here in XGBoost is similar to what was done to obtain Figure 4.21 except the criterion (MSE versus Equation (4.7)), we simply explain how we would split a parent node to obtain child nodes *left* and *right*. We simply make a split at the feature value that maximizes the information gain (like before) but here that is given by

$$\frac{1}{2} \left[\frac{\left(\sum_{i \in L_{left}} g_i\right)^2}{\sum_{i \in L_{left}} h_i + \lambda} + \frac{\left(\sum_{i \in L_{right}} g_i\right)^2}{\sum_{i \in L_{right}} h_i + \lambda} - \frac{\left(\sum_{i \in L_{parent}} g_i\right)^2}{\sum_{i \in L_{parent}} h_i + \lambda} \right] - \gamma$$

which is obtained directly from Equation (4.7) by applying to the left and right leaves of a parent node. This iteratively built tree in a greedy fashion starting with all the samples as the root ends up as the bth tree. It is worthwhile noting that if we used $\lambda = 0$, $\gamma = 0$, and $\ell(\cdot,\cdot)$ as the squared error, then the approach becomes identical to what was used to obtain the tree in Figure 4.21.

In Figure 4.21 we used a brute-force approach by trying all the m features across all the potential splits through a complete enumeration. The XGBoost algorithm uses a faster approach to find the splits. We do not explain that in this book and the interested reader is referred to Chen and Guestrin [10] for details. One other aspect that is used in XGBoost is the notation of a learning rate η. This is not explained in Chen and Guestrin [10]. However, it is relatively straightforward. A learning rate η is pre-multiplied to the residuals before they are summed for the final prediction using the B trees

$$\hat{y}_i = \bar{y} + \eta \sum_{b=1}^{B} f_b(\mathbf{x}_i)$$

where $\bar{y}$ is the average of the y_i values and considered as the prediction for the zeroth tree, i.e., $\hat{y}_i^{(0)} = \bar{y}$. One typically uses a small learning rate η (such as 0.1) and a large number of trees B (such as 100) as they work together to prevent over-fitting and improve model performance.

Having described the building blocks of XGBoost, we now delve into the Python code. In the process we will also state what parameters need to be set. The first cell of the python code is identical to that in Figure 1.6. The second to eighth cells are somewhat similar to what was used in Gradient Boosting in Figures 4.1, 4.2, 4.3, and 4.4. However, there are some key differences. We do not restate the aspects that are similar and the reader is encouraged to refer to the description in Gradient Boosting. But we will go over the details of the aspects that are different. We start with cell-2 which is significantly different.

Cells 2 and 3 for the XGBoost implementation in Python are described in Figure 4.22. Line 2 in cell-2 imports the XGBoost regressor from the `xgboost` library. Lines 3 and 4 of cell-2 are mainly used to tune the parameters by splitting the data used to learn, searching through a space of values, and compute the mean squared error. We will describe those in cells 9 and 10 later. Although Cell-2 was different from what was described in Gradient Boosting, Cell-3 is almost identical. The only difference is that we used scaled irradiance data here. Lines 3 and 4 of Cell-3 retrieve 3 months of scaled solar irradiance for the years 2016 and 2017. Notice the scaled global irradiance values in the outputs of Cell-3.

In Figure 4.23, Cells 4 and 5 are presented. They are nearly identical to Figures 4.2 and 4.3. The main difference is in the various columns of data. This data set has one fewer month and so, the features are slightly different. Also, in Cell 5, it is important to notice that we only consider three lag features of scaled irradiance for the predicted time: 24 hours ago (`lags`), 25 hours ago (`lags-1`), and 48 hours ago (`lag2s`). This is simply to show how the method works. One can consider other lags as well such as one or two more periods ahead. Such issues of feature engineering are not presented here but would be important to consider while building a forecasting model. Cells 6, 7, and 8 are shown in Figure 4.24 which are nearly identical to Figure 4.4 except the weather conditions due to a month shorter is slightly different here.

```
[2]:
# Key imports for this code
from xgboost import XGBRegressor
from sklearn.model_selection import train_test_split, GridSearchCV
from sklearn.metrics import mean_squared_error
```

```
[3]:
seasonal_periods = 8
# fetch data
prev_year = pd.read_csv('Golden_hourly_2016_3_months_scaled.csv', parse_dates = ['datetime'])
solar_data = pd.read_csv('Golden_hourly_2017_3_months_scaled.csv', parse_dates = ['datetime'])
prev_year.head()
solar_data.head()
```

[3]:

	datetime	hour	scaled _global	Denver _temp	Denver _weather
0	2016-09-01 08 :00 :00	8	0.85	287.84	sky is clear
1	2016-09-01 09 :00 :00	9	0.88	287.65	sky is clear
2	2016-09-01 10 :00 :00	10	0.91	287.05	sky is clear
3	2016-09-01 11 :00 :00	11	0.92	286.64	sky is clear
4	2016-09-01 12 :00 :00	12	0.92	285.95	sky is clear

[3]:

	datetime	hour	scaled _global	Denver _temp	Denver _weather
0	2017-09-01 08 :00 :00	8	0.83	290.71	sky is clear
1	2017-09-01 09 :00 :00	9	0.74	289.78	sky is clear
2	2017-09-01 10 :00 :00	10	0.48	288.83	sky is clear
3	2017-09-01 11 :00 :00	11	0.89	288.37	scattered clouds
4	2017-09-01 12 :00 :00	12	0.90	288.48	sky is clear

Figure 4.22 Cells 2 and 3 of the Python Code for Day Ahead Forecasting with XGBoost

```
[4]:
input_data = pd.get_dummies(solar_data, drop_first=True)
input_data.columns
```

```
[4]:
Index(['datetime', 'hour', 'scaled_global', 'Denver_temp',
       'Denver_weather_few clouds', 'Denver_weather_fog',
       'Denver_weather_haze', 'Denver_weather_light rain',
       'Denver_weather_light snow', 'Denver_weather_mist',
       'Denver_weather_other', 'Denver_weather_overcast clouds',
       'Denver_weather_scattered clouds', 'Denver_weather_sky is clear'],
      dtype='object')
```

```
[5]:
input_data['lags'] = input_data['scaled_global'].shift(periods = seasonal_periods)
input_data['lags-1'] = input_data['scaled_global'].shift(periods = seasonal_periods+1)
input_data['lag2s'] = input_data['scaled_global'].shift(periods = 2*seasonal_periods)
input_data.dropna(inplace = True) # This will drop the first two full days
```

Figure 4.23 Cells 4 and 5 of the Python Code for Day Ahead Forecasting with XGBoost

Next we move to Cell-9 which is shown in Figure 4.25. The main purpose of this cell is to select the hyper-parameters used in the XGBoost model. The way this is done is to try a discrete set of parameter values and see which one performs best in that set. To do that, we create two data frames in lines 2 and 3 of Cell-9. Line 2 creates the *X* variables by ignoring the 'datetime' column and the *y* variable, and line 3 creates the *y* variables from the 2016 data. Then in line 5 we split *X* and *y* into two groups, one (80% of the data) to train various XGBoost models and other (remaining 20%) to test those models. The test data is called cross-validation data as

```
[6]: xgb_train_data = pd.get_dummies(prev_year, drop_first=True)
```

```
[7]: xgb_train_data['lags'] = xgb_train_data['scaled_global'].shift(periods = seasonal_periods)
     xgb_train_data['lags-1'] = xgb_train_data['scaled_global'].shift(periods = seasonal_periods+1)
     xgb_train_data['lag2s'] = xgb_train_data['scaled_global'].shift(periods = 2*seasonal_periods)
     xgb_train_data.dropna(inplace = True) # This will drop the first two full days
```

```
[8]: input_data['Denver_weather_other'] = input_data['Denver_weather_other'] + input_data['Denver_weather_light snow']
     input_data.drop(columns = ['Denver_weather_light snow'], inplace = True)
```

Figure 4.24 Cells 6, 7, and 8 of the Python Code for Day Ahead Forecasting with XGBoost

```
[9]:
1  # Creating training and cross validation data using previous year
2  X = xgb_train_data.drop(columns = ['datetime', 'scaled_global'])
3  y = xgb_train_data['scaled_global']
4  # Split into training and validation sets
5  X_train, X_val, y_train, y_val = train_test_split(X, y, test_size=0.2, random_state=42)
6  # Define the XGBoost regressor
7  model = XGBRegressor(objective='reg:squarederror', random_state=42)
8  # Hyperparameter grid
9  param_grid = {
10     'n_estimators': [50, 100, 400],      # Number of trees
11     'max_depth': [3, 5, 7],                  # Maximum depth of trees
12     'learning_rate': [0.05, 0.1, 0.3],       # Step size shrinkage for updates (eta)
13     'gamma': [0, 20.0, 50.0],                # Minimum loss reduction to make a split
14     'min_child_weight': [1.0, 5.0, 8.0],     # Minimum weight to split node into child
15     'reg_lambda': [1, 16, 32],               # Lambda used in weight feature
16     'colsample_bytree': [0.8, 1.0],          # Fraction of feature columns to choose
17     'subsample': [0.7, 1.0],                 # Fraction of sample rows to select
18 }
19 # GridSearchCV for hyperparameter tuning
20 grid_search = GridSearchCV(estimator=model, param_grid=param_grid, cv=3, scoring='neg_mean_squared_error', verbose=1)
21 grid_search.fit(X_train, y_train)
22 # Best model
23 best_model = grid_search.best_estimator_
24 grid_search.best_params_
```

Figure 4.25 Cell 9 of the Python Code for Day Ahead Forecasting with XGBoost

this is simply to select the hyper-parameters. This should not be confused with the final model testing. Line 7 defines the model and says that the criterion would be squared error (which is the $\ell(\cdot,\cdot)$ we considered earlier in this section).

There are several parameters to decide in XGBoost, namely, the maximum depth of the tree (`max_depth`), and conditions on whether to split a node into child nodes. In gradient boosting, for the latter we simply used the minimum number of samples in a leaf. However, here in XGBoost, since the leaf nodes have different weights w_l, we use a criterion that decides what would be too small to split (`min_child_weight`). Then we also need to set the learning rate η, the regularization for the weight λ, the regularization for the number of leaf nodes γ (it is also equal to the minimum loss reduction to make a split), and the number of trees B (`n_estimators`). In addition, it is also possible to select a fraction of the features (`colsample_bytree`) and the fraction of samples (`subsample`).

Line numbers 10 to 17 in Cell-9 described in Figure 4.25 show the various alternatives for the above hyper-parameters. They are presented in a different order than what is explained in the previous paragraph. However, from the name and the comments in the figure, the specific hyper-parameter should be easy to understand. Notice that we consider 2–3 options for each including some default values. The reason for only providing 2–3 options is that these take time to run. Giving more choices for some parameters that are more sensitive would be reasonable. Also, not all the

```
# Evaluate on in-sample set to check fit
rmse_train = np.sqrt(mean_squared_error(y_train, best_model.predict(X_train)))
print(f"Training RMSE: {rmse_train}")
# Evaluate on validation set to check overfit
y_pred_val = best_model.predict(X_val)
rmse_cv = np.sqrt(mean_squared_error(y_val, y_pred_val))
print(f"Validation RMSE: {rmse_cv}")
# Predict on test set
df_test = input_data.drop(columns = ['datetime', 'scaled_global'])
df_test_predictions = best_model.predict(df_test)
rmse_test = np.sqrt(mean_squared_error(input_data['scaled_global'], df_test_predictions))
print(f"Test Data RMSE: {rmse_test}")
# Save or return predictions
df_test['y_pred'] = df_test_predictions
```

```
Training RMSE: 0.18195464739545977
Validation RMSE: 0.20411478551959697
Test Data RMSE: 0.28816427447354037
```

```
# Lists to store the percentage and absolute errors
xgb_perc_error_list = []
xgb_abs_error_list = []
y_test = input_data['scaled_global']
# Make predictions
y_preds = df_test['y_pred']
# Calculate percentage and absolute errors
perc_errors = np.abs(y_test-y_preds)/y_test
abs_errors = np.abs(y_test-y_preds)
# Store the percentage and absolute errors
xgb_perc_error_list.append(perc_errors)
xgb_abs_error_list.append(abs_errors)
```

Figure 4.26 Cells 10 and 11 of the Python Code for Day Ahead Forecasting with XGBoost

hyper-parameters need to be selected, some can be at the default. This decision is not fixed and varies from code to code. Further, what values to input is also not obvious.

It would be a good idea to do this in multiple phases. In the first phase we can try out some arbitrary values. Then we can fine-tune with fewer or more values, and also trying values that are different. Once we are quite satisfied that the training model does not over-fit, we can finalize a set of parameters. The issue of over-fitting will be described later. For any set of values, the code essentially tries out by a complete enumeration of all combinations. This is described in line 20. It searches a "grid" of all possible values (in each of the 8 hyper-parameters). Then, in line 21 of Cell-9 it fits an XGBoost model and compares the mean-square error seen in the validation data set. Then in line 23 it shows the best set of parameters among those given, and line 24 asks to display the result.

Next refer to Figure 4.26 which describes cells 10 and 11 of the python code. Cell-10 essentially compares the root mean square error (RMSE) of the training data, the cross-validation data, and the testing data for the *best* hyper-parameters among those provided in the grid search in Cell-9. Recall that training is done across all the hyper-parameters in consideration using 80% of randomly selected rows in the 2016 data. Then the models are compared for the mean square error for the 20% data set aside. The best model results in the lowest sum of square error (and hence RMSE as well). The test data is the data from 2017 where we use the best model trained on all the training data of 2016.

[14]:

```
 1 # Creating training data
 2 X_train = xgb_train_data.drop(columns = ['datetime', 'scaled_global'])
 3 y_train = xgb_train_data['scaled_global']
 4 # Define the XGBoost regressor with specific hyperparameters
 5 model = XGBRegressor(
 6     n_estimators=1200,
 7     max_depth=8,
 8     learning_rate=0.005,
 9     min_child_weight = 25.0,
10     colsample_bytree=0.4,
11     gamma = 60.0,
12     objective='reg:absoluteerror',
13     random_state=42
14 )
15 # Train the model
16 model.fit(X_train, y_train)
```

Figure 4.27 Final Choice of Hyper Parameters for the Python Code with XGBoost

Notice that the RMSE does change significantly based on the output of Cell-10. That indicates that the model is perhaps over-fitting, i.e., there is variance. So we would perhaps need to find a different set of hyper-parameters. However, hyper-parameters may not be the only thing that is causing the over-fitting. We may also reconsider the features used. While forecasting for a single location, it is worthwhile fine-tuning the model and selecting hyper parameters multiple times before finalizing on a set that results in reasonable performance with little or no over-fitting. Finally, Cell-11 collects the performance measures for the test data to compute metrics we have used before, namely the percentage and absolute errors.

As a last figure in this section, consider the cell in Figure 4.27. Once the hyper-parameters are selected, we could use such a cell to replace the Cells 9 and 10 we presented earlier. Here we use the entire training set (and not have any cross validation data that is set aside for hyper-parameter tuning). The lines 2 and 3 of this cell essentially uses the 2016 data available and retrieves the right features to train and provide ground truth. Then lines 5 to 13 describe the hyper parameters and model set up values. Notice that none of the hyper parameters are from the original set used in Cell-9. Also notice two of the hyper-parameters are not considered (as we simply let them take the default values). Finally notice that the objective now is the absolute error which we wish to minimize. We would run such a code and produce the output. In the next section we compare the various models.

4.4 PERFORMANCE OF ALGORITHMS AGAINST PERSISTENT METHOD

The purpose of this section is to benchmark the performance of the various approaches seen thus far against the persistent forecast. Persistent forecast in a day-ahead sense is to forecast the scaled irradiance we are observing right now as the forecast for the same hour next day. Although this is a naive forecasting approach it is actually very powerful. Our goal is to see how the methods such as exponential smoothing, SARIMAX, gradient boosting (GB), and XGBoost compare against the persistent forecast. A few caveats are important to state. One is that the models have not been fine-tuned to select the best possible parameter values (the idea is to walk through the reader see where there is scope for improvement and what might one try).

The second caveat is that there is no guarantee that the methods that work well in this application would work well in all geographical locations for all horizons of forecasts using any set of exogenous features. It would be best if the reader tries out all the methods, carefully fine-tunes the parameters and model features, and then adopts what is best for their setting. The third caveat is that we only present the same eight metrics we saw in the previous chapter, namely, the mean, median, 75th percentile, and 90th percentile of both absolute percentage error (APE) as well as absolute error ratio (AER). Recall that the latter is the ratio of the absolute error and the mean of the ground truth in the test data. Improving one metric could worsen another metric. Also, metrics such as RMSE have not been considered.

Although we have mentioned the data sets and the scaling earlier in this chapter, it is worthwhile restating for the same of completeness. The data is from NREL [34] and it includes global horizontal irradiance which is scaled to a number between 0 and little over 1. The process of scaling is by accounting for the rotation, tilt, and revolution of the earth around the sun. We consider the same three locations as the ones in the previous chapter. The first site is called SolarTAC in Aurora in the state of Colorado. Its latitude is 39.75685°North and longitude is 104.62025°West. The elevation is 1674 meters above mean sea level. The data is collected at a granularity of 1 minute (for 3 months in 2016 and 2017) and averaged over an hour.

A similar dataset is obtained from the site SRRL in Golden, Colorado (CO). Its latitude is 39.742°North and longitude is 105.18°West, with an elevation of 1828.8 meters above mean sea level. The data is collected during the same time period and process as in Aurora, CO. The third dataset is from Los Angeles from the site SOL-RMAP located in Loyola Marymount University. Its latitude is 33.966674°North and longitude is 118.4226°West with an elevation of 27 meters above mean sea level. Although the granularity of the data is similar, here the months and years are different. They are collected during the winter months of 2012-13 and 2013-14, 1.5 months in the calendar later than those collected in Aurora and Golden.

BASE MODELS

In Table 4.4 we present the results for the aforementioned metrics across the three locations using five base models. These base models are exponential smoothing (ES), SARIMAX, Gradient Boosting (GB), XGBoost, and Persistent Forecast. For the ES and SARIMAX models, we used a training data of 1 week and each new hour the models were retrained to make a prediction $h = 8$ periods later. The seasonality was also 8 (which is the number of sunlight hours in a day during winter). For ES an additive trend and a multiplicative seasonality were used. For SARIMAX the orders were $(p,d,q) = (2,1,0)$ and $(P,D,Q) = (0,0,1)$. Also, for SARIMAX the weather

Table 4.4
Error metrics for 24-hour ahead forecasts in multiple cities for 1-hour intervals

Location	Method	Metric	Mean	Median	75th perc.	90th perc.
Aurora, CO	Exp. Smoothing	APE	55.2%	20.0%	58.0%	111.3%
		AER	31.7%	20.6%	48.3%	77.7%
	SARIMAX	APE	57.1%	21.9%	63.3%	131.0%
		AER	33.6%	22.7%	52.5%	82.8%
	GB	APE	53.8%	12.2%	38.2%	132.5%
		AER	24.1%	13.4%	36.0%	63.2%
	XGBoost	APE	53.7%	11.6%	36.9%	137.8%
		AER	23.5%	13.7%	35.3%	62.7%
	Persistent fcst.	APE	52.9%	17.3%	55.9%	115.1%
		AER	29.8%	17.2%	48.5%	76.8%
Golden, CO	Exp. Smoothing	APE	97.1%	40.0%	86.8%	210.9%
		AER	44.9%	33.7%	68.2%	103.5%
	SARIMAX	APE	95.2%	38.8%	75.1%	215.5%
		AER	42.4%	31.4%	66.5%	93.7%
	GB	APE	106.3%	21.4%	74.0%	291.7%
		AER	34.3%	24.1%	54.0%	81.6%
	XGBoost	APE	104.4%	18.3%	69.3%	278.3%
		AER	33.0%	21.4%	51.8%	78.2%
	Persistent fcst.	APE	95.7%	39.8%	76.6%	225.6%
		AER	42.3%	32.4%	68.8%	98.6%
Los Angeles	Exp. Smoothing	APE	32.6%	12.6%	41.3%	74.9%
		AER	23.4%	12.7%	33.9%	59.7%
	SARIMAX	APE	32.9%	14.0%	39.5%	80.1%
		AER	24.2%	14.2%	35.5%	61.8%
	GB	APE	30.2%	16.6%	29.7%	53.8%
		AER	21.2%	17.3%	31.1%	43.8%
	XGBoost	APE	29.4%	16.5%	26.8%	50.1%
		AER	20.6%	16.8%	28.0%	40.1%
	Persistent fcst.	APE	29.6%	9.6%	36.8%	70.3%
		AER	20.6%	10.1%	32.4%	56.4%

features and temperature forecasts were used. For all the sites, the forecast quality is only comparable with persistent as evident from Table 4.4. ES and SARIMAX become inaccurate several periods out and they are usually more effective when the horizon is 1 or 2.

For GB, the models used 100 trees (`n_estimators`), maximum depth of 5, and minimum samples in each leaf of 2. For XGBoost a grid search was performed with: 50, 100, and 200 trees; 3, 5, and 7 as maximum depth; learning rate of 0.01, 0.1, and 0.2; sub-sample of 0.8 and 1.0; and fraction of features as 0.8 and 1.0 for all the locations. The best parameter sets were the same across the board except the fraction of columns were 0.8 for Golden and Los Angeles but 1.0 for Aurora. For the two Colorado locations Aurora and Golden, both GB and XGBoost for the mean and median metrics performed better than persistent forecast. However, for Los Angeles, although the errors themselves were extremely low, they were not better than persistent for any of the methods. This requires a deep dive to see what causes the issue.

As a first deep dive, we could try and verify if there is over-fitting in case of GB. A way to check that is to see if the in-sample predictions are similar in accuracy as the out-of-sample data. In fact, the training data had a smaller error than the testing data indicating that there is over-fitting in the training process. So a different set of parameters were used, in particular 50 trees and minimum samples of 3 in each leaf. That resulted in APE results of 24.6%, 8.4%, 19.1%, and 57.2% for mean, median, 50th percentile, and 75th percentile, respectively. These numbers were somewhat closer to the testing data prediction. Likewise the AER percentages were 11.7%, 7.9%, 15.8%, and 27.2% respectively for mean, median, 50th percentile, and 75th percentile. These values for Los Angeles are significantly better than the persistent forecast values.

Next, for Los Angeles, we try XGBoost with different parameters than those used in the earlier best prediction. In particular, `max_depth = 4`, `min_child_weight = 5`, and `gamma = 50` with the other parameters same as earlier. For that, the testing data produces APE values of 29.3%, 10.2%, 22.6%, and 63.9% for the mean, median, 50th percentile, and 75th percentile, respectively. Likewise, the numbers for AER percentages were 17.4%, 10.6%, 24.6%, and 41.1% for the mean, median, 50th percentile, and 75th percentile, respectively. These numbers are also yield much better results than persistent forecast. There are also other ways to improve the model accuracy. One of that is to use hybrid models which we consider next.

HYBRID MODELS

Here we present two ways of building hybrid models in the hopes of obtaining more accurate forecasts. The first one is fairly straightforward where we use the predictions of one model as a "feature" for the other models. As an example let us consider the ES predictions as a feature for the other models. One reason to use ES is that it does not use any exogenous variables. However, it is adept at handling seasonality and trends. Then we use the ES predictions as an additional feature for SARIMAX, GB, and XGBoost. The results are presented in Table 4.5. They are made for the same three locations using the same data and parameters as those used to obtain the results in Table 4.4. Clearly, persistent forecasts results are the same as it is the same dataset. The other methods show small improvements but not much.

It is important to note that for the training data, the predicted ES data is used as features and not the ground truth (of course, for testing that is the only option). But the small improvement must not be too surprising as the models already are over-fitting and we are adding a feature to that. Since the ES model is handling seasonality and trends, it may not be a bad idea to not consider the lag features at all, or at least drop a few of them. Those results are not shown here but would be good ones to consider as an alternative for model improvement. Again, it would also

Table 4.5
Error metrics day ahead forecasts using ES as feature in multiple cities for 1-hour intervals

Location	Method	Metric	Mean	Median	75th perc.	90th perc.
Aurora, CO	SARIMAX w/ ES	APE	56.5%	22.7%	61.0%	127.1%
		AER	33.7%	23.2%	50.8%	80.5%
	GB w/ ES	APE	53.9%	11.1%	39.2%	130.2%
		AER	23.8%	12.8%	36.1%	62.8%
	XGBoost w/ ES	APE	53.8%	11.7%	37.0%	139.8%
		AER	23.5%	13.8%	34.9%	61.3%
	Persistent fcst.	APE	52.9%	17.3%	55.9%	115.1%
		AER	29.8%	17.2%	48.5%	76.8%
Golden, CO	SARIMAX w/ ES	APE	95.5%	38.3%	77.6%	228.8%
		AER	42.8%	32.2%	65.6%	95.9%
	GB w/ ES	APE	105.1%	20.3%	71.6%	293.1%
		AER	33.8%	22.8%	52.6%	84.2%
	XGBoost w/ ES	105.3%	18.4%	64.6%	282.4%	
		AER	32.6%	21.2%	50.5%	79.6%
	Persistent fcst.	APE	95.7%	39.8%	76.6%	225.6%
		AER	42.3%	32.4%	68.8%	98.6%
Los Angeles	SARIMAX w/ ES	APE	33.5%	14.2%	40.4%	80.4%
		AER	24.7%	14.4%	35.3%	61.1%
	GB w/ ES	APE	30.4%	16.0%	29.3%	56.6%
		AER	21.3%	16.7%	29.9%	44.1%
	XGBoost w/ ES	APE	29.3%	16.6%	26.8%	52.8%
		AER	20.5%	17.3%	27.6%	39.0%
	Persistent fcst.	APE	29.6%	9.6%	36.8%	70.3%
		AER	20.6%	10.1%	32.4%	56.4%

benefit by trying different hyper-parameter values to improve the results. However, let us switch to a second hybrid model now.

The second hybrid model is a bit nuanced. The key idea is that one model may perform well under one set of input values while another might perform better at others. A simple way to address that is to simply take the average of multiple forecasts. But a more involved approach uses residuals. Let us say there are two models, A and B. Let $\overline{y}_{train}$ and X_{train} be the training sample's ground truth and feature set, respectively. Say that model A produces training data predictions as $\hat{y}^{A}_{train}$ and testing data predictions as $\hat{y}^{A}_{test}$. Then model B is trained on the residuals, i.e., $\overline{y}_{train} - \hat{y}^{A}_{train}$ is used as the ground truth and with X_{train}, say that it produces residual predictions as $\hat{y}^{B}_{test}$ (realizing this is an estimate of the residual). Then we predict $\overline{y}_{test}$ as the sum $\hat{y}^{B}_{test} + \hat{y}^{A}_{test}$.

We build such hybrid models and present the results in Table 4.6. As the initial model (the model A in the previous paragraph) we consider a Random Forest (RF) model. One could have again used ES but we wanted to display the results

Table 4.6
Error metrics day ahead forecasts using residuals from random forest in hybrid methods for 1-hour intervals

Location	Method	Metric	Mean	Median	75th perc.	90th perc.
Aurora, CO	Random Forest	APE	51.9%	11.6%	37.7%	126.9%
		AER	23.1%	13.8%	34.3%	59.2%
	GB w/ RF	APE	53.4%	13.8%	39.6%	126.5%
		AER	24.7%	15.6%	38.0%	62.3%
	XGBoost w/ RF	APE	54.0%	10.5%	42.1%	135.6%
		AER	23.4%	12.3%	34.7%	62.9%
	Persistent fcst.	APE	52.9%	17.3%	55.9%	115.1%
		AER	29.8%	17.2%	48.5%	76.8%
Golden, CO	Random Forest	APE	106.2%	17.6%	75.2%	292.5%
		AER	32.8%	21.4%	48.4%	80.4%
	GB w/ RF	APE	101.9%	21.8%	70.8%	283.1%
		AER	33.9%	24.2%	53.2%	81.4%
	XGBoost w/ RF	APE	117.3%	15.9%	92.0%	331.1%
		AER	33.4%	19.1%	56.2%	90.2%
	Persistent fcst.	APE	95.7%	39.8%	76.6%	225.6%
		AER	42.3%	32.4%	68.8%	98.6%
Los Angeles	Random Forest	APE	28.8%	15.4%	23.9%	55.0%
		AER	19.4%	16.2%	25.8%	36.5%
	GB w/ RF	APE	29.6%	16.9%	28.7%	53.4%
		AER	20.7%	17.3%	27.6%	44.0%
	XGBoost w/ RF	APE	28.6%	10.7%	21.3%	64.8%
		AER	17.2%	11.8%	22.5%	40.4%
	Persistent fcst.	APE	29.6%	9.6%	36.8%	70.3%
		AER	20.6%	10.1%	32.4%	56.4%

for a different initial model. Also note that RF works in a different manner than GB and XGBoost (that are somewhat similar). Hence there is a possibility that RF may perform well in different values of the feature space than GB or XGBoost. The parameter used for RF in all the three cities were 100 trees, maximum depth of 5, and minimum samples in leaf as 2. They were also the exact same parameters for GB. The parameters used in XGBoost were `n_estimators = 50`, `max_depth = 4`, `learning_rate = 0.1`, `subsample = 1.0`, `min_child_weight = 5.0`, `gamma = 50.0`, and `objective='reg:absoluteerror'` with a random state of 42.

The above parameter values were set the same in all cities here for the ease of presentation. In practice, it is recommended that each location be fine-tuned with the right set of parameters. Refer to Table 4.6. For the sake of completeness, the RF results are shown. It is in-fact quite competitive in terms of quality. It does outperform the GB model although we must note that the GB model is a hybrid that uses the residuals from the RF model. It is entirely possible that there is some amount of over-fitting. Even the XGBoost model does not get a tremendous improvement although it, like GB, also uses a hybrid model trained on the residuals from the RF model. These results are only a minor improvement from the previous results. The XGBoost models actually use a different set of parameters that could explain the quality of the results. Although the Aurora and Golden results for persistent are easily beaten, once again for the Los Angeles data, they are comparable. This prompts us to consider further investigation which is done next.

FURTHER INVESTIGATIONS

The purpose of this further investigation of the Los Angeles predictions is to demonstrate how one would go about finding ways to improve the model. It is crucial to realize that there are two things that need improvement, one is to reduce (or remove) over-fitting while the other is to improve both the training and testing predictions. We saw earlier that there is a significant amount of over-fitting so that the training data errors are tiny but when tried on the test data they do not perform as well. It is not that the sophisticated model is no match for the simple persistent model, it does beat it in the training data, and in fact the errors are small to begin with. The question to ask is under such a situation what does one do.

In our earlier analysis we saw that the weather did not really play a major factor and was perhaps more of a distraction in the hybrid model. So we drop weather as a feature, however, we add the month as a feature thinking that the three months (November, December, and January are quite different). One can make a case for further dividing into half months but that was not done. As a first part of the hybrid approach, we fit an RF model. This time, instead of using the residuals, we simply use the RF predictions as a feature for the XGBoost model. We search through a set of hyper-parameters. Unfortunately all of them severely over-fit. So we considered a different set of parameters for XGBoost.

The final parameter set that we used for XGBoost were `n_estimators = 1200`, `max_depth = 8`, `learning_rate = 0.005`, `subsample = 0.8`, `gamma = 20.0`, `min_child_weight = 25.0`, `colsample_bytree = 0.4`, `reg_lambda = 80.0`, and `objective='reg:absoluteerror'` with a random state of 42. Except for the last two, all the hyper-parameters are significantly different from what was used before and also way outside the values considered in the grid search. It produces APE metrics of 27.9%, 8.98%, 21.5%, and 66.3% for the mean, median, 75th percentile, and 90th percentile, respectively. Likewise, the AER values were 16.5%, 9.84%. 22.0%, and 43.7% for the mean, median, 75th percentile, and 90th percentile, respectively.

There are a few things to note. Firstly, every metric is better than the corresponding ones for the persistent forecast. Although that is nothing to celebrate as the persistent method is an extremely simple approach. However, two main issues emerge: the training data has a much worse results than before although the testing data has vastly improved. So what we have done is to reduce the over-fit problem. Although it must be stated that even this model is in some degree over-fit. Secondly, the features compared to the previous over-fit models are different. As stated in the previous

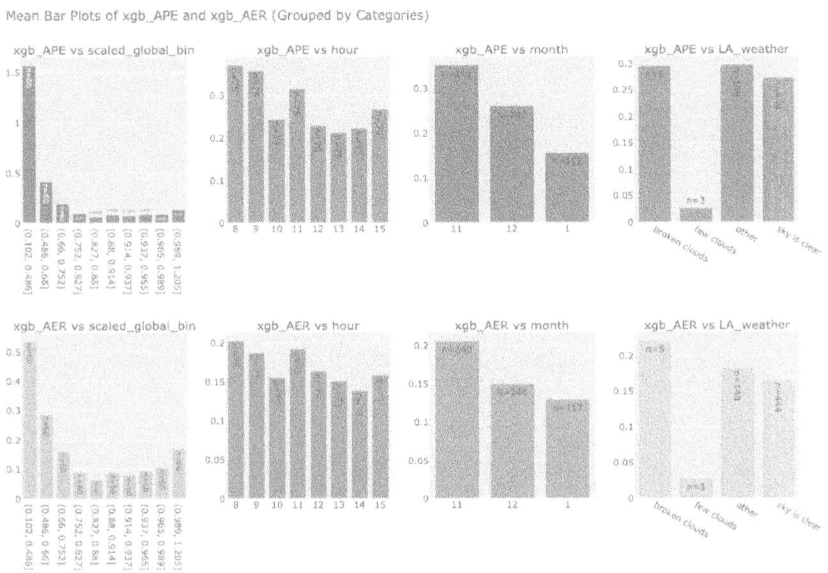

Figure 4.28 Bar plot of grouped average performance and number of samples for Los Angeles

paragraph, we have removed the weather feature and added month. Thirdly, we have used the RF prediction as a feature and not through residuals.

In Figures 4.28 and 4.29 we investigate further to see if the errors are worse in certain circumstances and not in others. For that, we look at four aspects: the ground truth (`scaled_global`), the hour of the day, the month of the year, and the type of weather. We explain one-by-one next but before that say what the two figures are. Figure 4.28 is a bar graph of the mean APE (top row) and mean AER (bottom row) for various values of the four aspects. By grouping the results this way, it is possible to see if the errors are higher for certain conditions. The figure also lists the number of samples "$n =$" to further understand the importance.

Figure 4.29 is a box plot of the same set with APE on the top and AER on the bottom, however, the box plots represent the spread including percentiles. The horizontal line inside each box shows the median value. The top of each box is the 75th percentile while the bottom of the box is the 25th percentile. Then there are two whiskers (for that reason these are also known as box and whisker plots) that depict the "maximum" and "minimum" values. The reason they are in quotes is because they are not the true maximum and minimum (but about 1.5 times the inter-quartile length) from the median. Points that are outside these whiskers are considered as outliers. Notice that all the points that are outliers are above and that is because the APE and AER are lower-bounded by 0.

Now let us analyze the four aspects (i.e., the four columns) in Figures 4.28 and 4.29. The motivation for the ground truth is to see if the quality of the results are

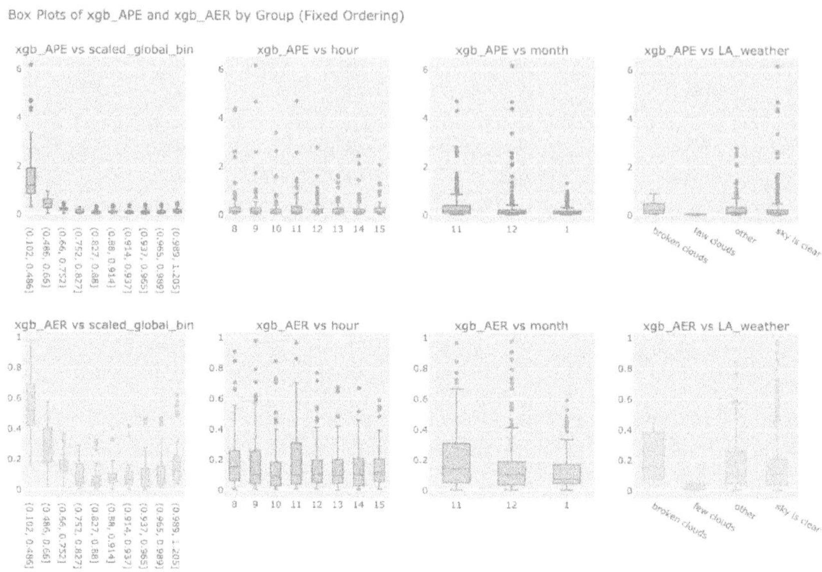

Figure 4.29 Box plot of grouped average performance indicating spread of metrics for Los Angeles

sensitive to the ground truth values, and they are indeed. For the APE, that is expected as the denominator is the ground truth, if that is small, even a small absolute error would get magnified. However, the AER also shows that the error is bad for small values of ground truth. That is because the absolute error itself has a good chance of being high. However, it is harder to explain why the errors are high when the ground truth is high. One possible explanation is that we did not consider weather predictions and when the number of clear-sky days are large (which is the case here), knowing that is useful.

The good news is that overcast days and clear sky days are easy to predict, especially in this day and age (the dataset we used does not have very good weather predictions and it also not for the exact sample point). So during the most difficult-to-predict cloud cover conditions our models do a great job of predicting. Notice that the buckets of ground truth are such that each of the 10 buckets has 60 data points. Next we move to the second column which is the hour of the day. Here, naturally, there are equal number of each of the hours. The APE and AER results for the mean in Figure 4.28 are nearly identical in shape. If we did not scale, there would have been a tremendous difference for the performance across hours. Scaling does allow the normalizing and there is no explainable difference among the hours even in Figure 4.29.

The third column is the plot performance against months. This gives us good reasons to have considered month as a feature as it seems to have an effect. In fact, for the RF model, month was more important a feature than hour, temperature, and 2-day lag. Although, for GB it was one of the least important, but that could be because the effects have been accounted for in the RF feature. Notice that the number of data points in January is less than half the other two months. The median values in Figure 4.29 are nearly similar, and the outliers possibly caused the major shift in means in Figure 4.28. One option is to remove the month feature from the second part of the hybrid model.

Finally, the last column is the LA weather related performance. Notice that nearly 3/4ths of the time the sky is clear (as a prediction, these are not actual, and also not at the specific location). Since there are very few samples for the first two, i.e., broken clouds and few clouds, it would not be a good idea to read too much into it. However both Figures 4.28 and 4.29 indicate that when the sky is clear the predictions are better. The fact that this is different from the first column leads us to believe that the forecasts of clear sky is not as accurate.

CONCLUDING REMARKS AND NEXT STEPS

That said we wish to conclude this section and chapter stating that what we have shown is just a sample of what can be done. Also, we have not found the best predictions. It can be improved by parameter tuning, including or removing various features (in fact, we did not consider fewer or additional lag features). One could also easily consider pressure information that can be nicely predicted. The key message is to try and improve the training data performance while keeping it nearly the same as the testing data performance. Also, only one metric should be used as improving a metric might worsen another. Before performing a study one must select the metric that would be most ideal (including things like RMSE that we did not consider).

Before moving to the next topic, it is worthwhile mentioning that in the next topic we would forecast what is known as the "net load". This is the demand experienced *minus* the solar supply. Many researchers simply use that instead of worrying about hours at night when there is no solar. By moving to net-load we could simply consider an entire day's power consumption. The methods are similar and so it would not be explained again in the next chapter. However, it is worthwhile pointing to some of the literature in that area (although this is by no means even close to the vast body of knowledge). We present these articles here and then directly jump into the problem in the next chapter.

We also talk about microgrids and smart buildings. A useful survey article to look into those would be Hernandez et al [16]. Hourly net load with PV through clear-sky and regression-based demand can be found in Peng et al [36]. Net load forecasting has used neural network models quite a bit and Stratman et al [42] consider behind the meter PV for that through LSTM (long short-term memory) models. Zhang et al [58] also use an ensemble of LSTM and other methods to accurately predict short-term net load for high PV penetration. Zhang et al [57] combine graph neural networks with transformers to yield net load forecasts. Also, Zhou and Wang [59] predict net load for behind-the-meter PV by considering a unified model using graph smoothing and spatio-temporal effects.

There are plenty of recent articles that discuss the recent trends in load-forecasting using LSTM (Li et al [31]), fuzzy ensemble models (Sideratos et al [39]), and wavelet transform (Alipour et al [1]) that are worth looking into as alternate approaches. However, there are also some articles from a few years ago that discuss short-term load forecasting (Choueiki et al [11] and Hippert et al [17]) that are worth looking into. There are many review articles such as Hou et al [18] discussing various challenges in load forecasting. Also, Wang et al [50] discuss probabilistic forecasting which we would get into later in the next chapter. However, it is also important to consider Shwartz-Ziv and Armon [38] who question if it is needed to use neural networks for tabular data.

5 Day-Ahead Planning

All other chapters in this book are on solar irradiance, time-series forecasting, or time-series forecasting for solar irradiance. This chapter, however, puts solar irradiance forecasting in the context of power system operations. In particular, we see how to manage a power system given the forecasts for planning a day in advance. Hence this chapter applies the results of the previous two chapters of day-ahead forecasts and short-term forecasts. However, those forecasts are point forecasts and in this chapter we also use distributional forecasts which we assume are known. Details of how to obtain distributional forecasts are in the next chapter. In summary, this chapter addresses the question: if we have forecasts of solar irradiance, how do we use them effectively?

We present a simplified setting in this chapter by abstracting away several of the complications found in real power system operations. The main idea is to understand in our hypothetical setting how one would use the forecasts to make planning decisions and then apply them to the realistic situations. The emphasis here is to manage the uncertainty in the solar forecasts, load (i.e., power demanded by electricity consumers such as homes, offices, and industry), and costs. We do not consider uncertainty in transmission and distribution or extreme events that strain the reliability of power systems. However, our hope is that by understanding the approaches presented, it would be possible for the reader to adopt the thinking to their systems by making suitable modifications.

An important aspect of this chapter is that the electricity price is *dynamic* and *stochastic*. Traditional electricity markets that use public utility commissions operated a regulated market. There the electricity prices are stable and consistent. However, we are moving toward widespread adoption of a deregulated market across the world where electricity generation is distributed and supplied by a different entity than the utilities that deliver it to the customers. In such a deregulated market, the price is heavily influenced by the supply, demand, and also other factors. As a result, the price of electricity changes randomly over time and space. The focus of this chapter is somewhat *futuristic* in that sense when it comes to price volatility.

In fact, we consider microgrid settings which are nearly-closed systems with their own solar power and diesel generators as well as their internal demand load. We assume that the microgrids participate in the energy markets by committing for power from the grid a day in advance and adjusting that in real time. A key assumption is that our microgrids are price takers and not price setters, and they pay for the electricity from the main grid (which will occur if the internal generation cannot meet the load). In Section 5.1 we present detailed description of a microgrid setting and explain the costs and objectives. Then, in Section 5.2, we consider the case when the microgrid does not have a storage for energy such as a battery (due to space or cost limitations). In Section 5.3, we consider models that allow for storage. In Section 5.4, we conclude the chapter by making a case for distributional forecasts.

DOI: 10.1201/9781003403500-5

5.1 SETTING, COSTS, AND OBJECTIVE

In this chapter we consider a setting of a microgrid in a loose sense that can easily be adapted to other situations that will be explained subsequently. According to U.S. Department of Energy (USDOE), "a microgrid is a group of interconnected loads and distributed energy resources within clearly defined electrical boundaries that acts as a single controllable entity with respect to the grid". An example of a single entity can be thought of as a campus such as a university, a company, or a housing community/neighborhood. This campus can run a microgrid which means that its electricity needs (also known as demand or load) are met through some locally distributed energy sources such as rooftop PV panels, wind turbines, diesel generators, microturbines, and fuel cells.

To meet the needs of the load demanded, microgrids can also buy power from the utility grid. When the supply of power from the distributed energy sources exceeds the power demand, microgrids can also sell power to the utility grid. There are also energy storage devices (viz., flywheels, batteries) in a microgrid as a buffer to combat the uncertainty and variability in the power demand and renewable supply. Now we parse through all the aspects of that USDOE definition in the previous paragraph. By using a campus framework, one can club all the loads (homes, offices, facilities, and infrastructure withing the campus) as interconnected loads and they are satisfied by a single entity that controls and manages power. The power itself is from locally distributed sources, and is aggregated and managed by interacting with the utility grid. Figure 5.1 illustrates the definition of a microgrid.

There are several important aspects to consider in the design and operation of a microgrid. However, the purpose of this chapter is not to go into those details. The main idea is to simply motivate that with the advent of solar PV generation on rooftops and other small-scale distributed locations, it is envisioned that a microgrid-like operation is a likely direction in the future. That said, in our system we consider a much simplified version than what is in Figure 5.1. We only consider solar PV as an

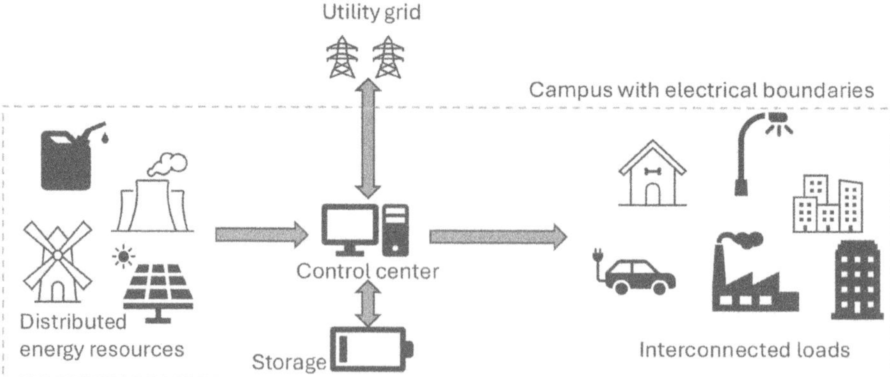

Figure 5.1 Abstract Notion of a Microgrid

internal renewable source and a large battery for storage. In addition, we consider a single aggregated load without going into specific details. In some sense, we consider a network of 5 nodes: solar PV source, battery, demand load, utility grid, and control center.

Also, one can easily think of special cases of the above 5-node network. For example, the case of load only or supply only or batteries only would all be special cases of what we will see in this chapter. The end goal is for the control center to balance the power supply and demand. That is the power (in kW) at every instant of time demanded is equal to the power (in kW) supplied. Of course, the control center is just a trans-shipment node that neither produces nor consumes power in our modeling abstraction. Also, the solar PV is only a supplier of power while the load center is only a demand of power. However, the battery and the utility grid can either give or receive power (if possible).

A key challenge in microgrids is uncertainty in the demand and supply. Due to the load profile (smaller number of loads than a typical utility), there is a lot of uncertainty and variability than a utility which enjoys the benefits that come with large amount of aggregation. Also there is uncertainty in the weather. Since there is tremendous localization, there is also a large amount of correlation. A cloud cover could potentially affect the entire microgrid. So on the one hand the variability and uncertainty is high, while on the other hand we may be able to consider a single forecasting model. A microgrid with time-varying and stochastic aggregate *demand* for power where a large fraction of demand met by solar energy sources (with *supply* which is time-varying and stochastic) situated locally and aggregated will be the setting we will consider.

Another uncertainty and time-varying aspect is the cost. This requires some explanation as this is more futuristic. We consider two costs one day-ahead and the other real-time that the utility grid charges the microgrid. By day-ahead we mean the cost per kWh for power to be committed from the utility grid used 24-hours from now. However, real-time means the cost per kWh the power would cost if we were to consume it right now (in the next 15 minutes). Notice the parallel between the above and what was mentioned in the short term and day-ahead forecasts in the previous two chapters. In the regulated market there is no such cost, however, there is such a cost in the deregulated markets. It is important to state that in the present moment these costs are not usually paid by microgrids to utility centers. We envision an operation of such a cost structure in the near future.

The rationale for such a cost structure is that the microgrid is expected to commit to certain amount of power a day in advance (this could be for buying or selling). This way the utility grid can plan its operations a day in advance. Then, in real-time due to the inherent uncertainties, the supply may go above or below the planned demand. Then one needs to buy or sell power in the real-time market. To make such commitment decisions a day ahead, one useful tool is to have a good forecast of the demand and supply. In this chapter we assume that such a forecast is available probabilistically. For now, we assume that we have a cumulative distribution function for the demand and supply forecasts.

We formalize the thought process in the above paragraph through some notations, assumptions, and definitions. A minor complication is that the day-ahead planning is at a 1-hour granularity while that of real-time is at a 5 or 15 minute granularity. To address that we use a 2-tuple notation for time (t,k) which means hour t from the start of the dataset, and k^{th} interval in that hour. To know the day index we can simply do $\lfloor t/24 \rfloor$ and to get the hour we could do $(t \mod 24)$. So $(419,3)$ would mean the 17th day in the dataset (starting with day 0), 11th hour (using a 24-hour clock), and third period (i.e., 11:30–11:45am). We do assume that the dataset begins with the first hour of the day. The reason for not simply using a 1-tuple is that in the day-ahead case we work with hourly granularity (which is the bulk of this chapter).

Let $\mathscr{A}$ be a set of 2-tuples of all possible values of (t,k) time-intervals in the dataset we will use for the analysis. To explain aspects of forecasting and what data will be used, we will consider notation of the type $\hat{y}_{(t,k)|(t-168,k)}$. That means we will forecast $y_{(t,k)}$ by training a model that uses all the data up until a week prior to (t,k) since $24 \times 7 = 168$. Although that is similar in flavor to what we saw earlier, the only addition here is that the estimate itself can be a random variable. Recall the 5 nodes in our network, except for the control center and storage (i.e., battery), the other three nodes have associated random quantities in each interval. We state those first.

For all $(t,k) \in \mathscr{A}$, let $D_{(t,k)}$ be the demand or load experienced at the microgrid during that interval. One can think of units of $D_{(t,k)}$ as kW (kilo-Watts) and this quantity is non-negative. Likewise, we let the solar power supply in that time interval from the PV source be $S_{(t,k)}$. We assume that we can convert irradiance that was forecasted into units of kW once we know the square-footage of the panels. The cost charged by the utility grid has two aspects one day-ahead $C^{DA}_{(t,k)}$ and other real-time $C^{RT}_{(t,k)}$. The units for both costs are \$ per kWh. However, the day-ahead costs are known a day in advance but the real-time costs are known only at the start of the interval (t,k) (so are $D_{(t,k)}$ and $S_{(t,k)}$).

Similar to cost, we also get a credit $B^{DA}_{(t,k)}$ and $B^{RT}_{(t,k)}$ for selling power back to the utility grid. Next we move to the fourth node of the 5-node network which is the storage (we call it battery or more generically ESD which stands for Energy Storage Device). Let $U_{(t,k)}$ be the amount of energy stored in the battery at the *beginning* of interval (t,k). Notice that unlike all the other notations, this definition applies to the beginning of an interval while all others stay constant across that interval. The units of $U_{(t,k)}$ is kWh which is a unit of energy, not power. Note that $U_{(t,k)}$ is a derived unit which would depend on the decisions or actions that we will describe next, hence it would not be found in the dataset.

In every time interval (t,k) we need to determine three quantities (which we call as decisions or actions). The first variable is $X_{(t,k)}$ which is the amount of power in kW we would *buy* from the utility grid. If the number if positive, it would correspond to the additional supply from utility grid to meet the demand in that interval. If it is negative, then it means there is excess supply that can be sold to the utility grid. The second variable is $Y_{(t,k)}$ which is the amount of power drawn from the solar PV cells during that interval in kW. The quantity $Y_{(t,k)}$ is non-negative. The third variable is $Z_{(t,k)}$ which is the supply in kW from the storage device, a positive value denotes discharging the battery so it acts as a source of power while a negative value means the battery is charging.

Next we state the objective. For that we specifically split the dataset into two pieces $\mathscr{A}_{train}$ and $\mathscr{A}_{test}$ so that the test set corresponds to the latter portion of $\mathscr{A}$ when arranged chronologically. We state the objective through a remark.

Remark

Objective: For all $(t,k) \in \mathscr{A}_{test}$, compute $X_{(t,k)}$, $Y_{(t,k)}$, and $Z_{(t,k)}$ so that the long-run average cost per hour is minimized. In the subsequent section we would explain when these decisions are made and with what information. We also state various constraints in terms of feasibility and capacity that need to be satisfied.

5.2 SYSTEM WITHOUT STORAGE: NEWSVENDOR MODEL

In the previous section we alluded to the day ahead versus real-time power management for microgrids. Here we begin by first motivating the need for such a process and then we go into the specific system which is a microgrid without an ESD. For the motivation we need to consider the issues faced by the utility grid. The utility grid needs to respond to the tremendous variability in both demand and supply. However, generators take a long time to get started if more need to be added to meet the demand (fossil fuel based generators are typically fastest but most others are slow). This prompts the utility grid to adopt a two time-scale pricing model.

The utility grid charges a day-ahead *price* (hourly) for power and the microgrid must make hourly commitments. The microgrid gets all the power they paid for during each hour of the day. There is also a real time *price* for power that the grid charges. At the microgrid, if there is a need for excess power (or some power needs to be sold back), the real time price kicks in. Recall from the previous section that the day-head price and credit are $C^{DA}_{(t,k)}$ and $B^{DA}_{(t,k)}$, respectively, while the real-time price and credit are $C^{RT}_{(t,k)}$ and $B^{RT}_{(t,k)}$, respectively, for all $(t,k) \in \mathscr{A}$. A day in advance, i.e., at $(t-24,k)$, we know deterministically what $C^{DA}_{(t,k)}$ and $B^{DA}_{(t,k)}$ are, and these are independent of k as the price and credit remain the same for the whole hour.

To give a concrete example, suppose that it is 2pm today. Then, for tomorrow's 2–3pm the microgrid can buy power from the utility grid at the rate of \$0.2 per kWh (or sell power back to the utility grid at the rate of \$0.1 per kWh). However, for tomorrow's 2–3pm the microgrid only knows a forecast of the load and a forecast of the solar power supply (both in kW, and both expressed as a probability distribution). The microgrid also has a forecast of the real-time price and credit for 2-3pm tomorrow, again probabilistically, such as a mean of \$0.5 per kWh to buy with standard deviation \$0.2 per kWh and a mean of \$0.05 per kWh to sell with standard deviation of \$0.01 per kWh. At 2pm today, the microgrid needs to commit a certain quantity of power, say $Q_{(t,k)}$ kW, for tomorrow's 2–3pm where (t,k) denotes the four 15-minutes periods from 2 to 3pm tomorrow.

Once the microgrid has committed $Q_{(t,k)}$, that amount stays the same for all k in that hour t, that is because the day-ahead price and power commitment is hourly. Also, $Q_{(t,k)}$ can be positive or negative denoting power purchased or sold to the utility grid. From the costs and credits standpoint, there is a clear incentive for the microgrid to commit a day ahead because it is cheaper to buy and more rewarding to sell a day in advance. For the utility grid, this incentive would also allow it to plan a day in advance by powering on and off flexible energy sources. However, the question is: what should $Q_{(t,k)}$ be? If the microgrid commits a day ahead to too much (or too little) power than what it is needed in real time, the excess (or shortage of) power must be sold back (or bought) at the real-time credit (or price).

Although the power demand $D_{(t,k)}$, solar PV supply $S_{(t,k)}$, real-time price $C^{RT}_{(t,k)}$, and real-time credits $B^{RT}_{(t,k)}$ are uncertain a day ahead, they will all be revealed at the beginning of the interval (t,k). Upon revealing, and given $Q_{(t,k)}$, it is fairly straightforward to determine the variables $X_{(t,k)}$, $Y_{(t,k)}$, and $Z_{(t,k)}$ stated in the **objective** of the previous section. This is because in this section we are in a situation where there is no storage or ESD in the microgrid. The motivation for that is the price of storage is high and it needs a decent amount of space to be lodged. From a modeling standpoint the implication is that in each interval (t,k) all the power produced must be consumed. So each interval can be solved in isolation.

ASSUMPTIONS AND NOTATIONS

We restate the assumption for the sake of completeness that in this section we only consider the case where the microgrid does not possess an ESD. Hence $Z_{(t,k)} = 0$ as there is no battery to even consider this variable. For simplicity we assume that all the prices and credits are positive and hence $Y_{(t,k)} = S_{(t,k)}$ as we would never curtail solar power and sell all the power to the utility grid or use it to satisfy the microgrid load, at least partially. If we know $Q_{(t,k)}$, then we can compute $X_{(t,k)}$ as

$$X_{(t,k)} = D_{(t,k)} - S_{(t,k)} - Q_{(t,k)}$$

as the extra amount we would buy from the utility grid (or sell if it is negative) after trying to satisfy the demand using the solar PV supply and the day-ahead commitments.

We also assume that we need to make a commitment for just one hour a day in advance (although in practice this could be different where a full 24-hours of commitment is necessary, but that is an easy exercise and is avoided to keep the notation more manageable). Thus, the only decision to be made is $Q_{(t,k)}$ for all $(t,k) \in \mathscr{A}_{test}$. In fact, since we can obtain $Q_{(t,k)}$ for each (t,k) independent of the others, we simply drop the (t,k) subscript from our notations for the rest of this section for ease of parsing. Hence, we denote $D_{(t,k)}$, $S_{(t,k)}$, $C^{DA}_{(t,k)}$, $C^{RT}_{(t,k)}$, $B^{DA}_{(t,k)}$, and $B^{RT}_{(t,k)}$ as D, S, C^{DA}, C^{RT}, B^{DA}, and B^{RT} respectively with the understanding that the subscript (t,k) is implicit.

PROBLEM STATEMENT

Let D be a random variable denoting the power demand that would need to be satisfied during time interval (t,k). Likewise, let S be a random variable denoting the power supply from renewable sources during that same interval. **Decision**: How much power Q to commit from the utility grid for hour t from the day-ahead market? We use the notation q to denote the quantity in the interval (t,k) with the understanding that $Q_{(t,k)}$ which we denote as Q must be the same for all k but $q_{(t,k)}$ which we denote as q does not. Note that we commit to buying the power a day ahead of (t,k) and the random variables D, S, C^{RT}, and B^{RT} will be realized at the beginning of interval (t,k).

Recall that if we have committed to buying more power than we need, we can sell back the excess at the real-time credit B^{RT}. Likewise, if we have committed to buying less power than we need, we fulfill our demand by buying in the real-time market paying C^{RT} per kWh. We *assume* that the day ahead price equals the day ahead revenue. This is reasonable since the day ahead prices are decided based on a market value of producers and consumers. However, if that is not the case, one has to be extremely careful while solving the problem as there would be multiple cases to consider. So, we have $C^{DA} = B^{DA} = d$, a deterministically known quantity a day ahead of (t,k).

So, we let d be the day-ahead price per kWh and it is deterministic. Then, the real-time price per kWh is C^{RT} and it is random with $\mathbb{E}[C^{RT}] = r$. Likewise, the credit per kWh B^{RT} in real-time is random with $\mathbb{E}[B^{RT}] = c$ (with the understanding that this could also be zero). We *assume* that

$$c \leq d \leq r$$

for non-triviality. The above condition is an incentive to commit a day ahead. For example, if $d > r$, then one could wait and simply buy in the real-time market. In a similar fashion, the inequality for selling also is an incentive to commit a day-ahead.

The decision of how-much-to-commit is repeated every 15 minutes every day, so minimizing the expected cost is a reasonable criterion. In fact, the objective in the previous section of minimizing the average cost per hour would be attained if in each interval (t,k) we minimize the average cost as there is nothing connecting from one interval to the next. The analysis and eventual decision is based on knowing the demand and supply characteristics (such as distributions) as well as costs' distributions. Recall that q is the quantity of power committed during interval (t,k) when the demand would be D and renewable supply S. So q is the decision variable that we wish to optimize.

This can be modeled in the same way a well-studied problem called the *newsvendor* problem is modeled and solved. For that reason, the title of this section has the word "newsvendor" in it. The topic of newsvendor is described in numerous texts and papers, one such reference is Petruzzi and Dada [37]. There the problem is to determine for a newsvendor the number of newspapers to buy and keep ready (i.e., stocked) early in the morning for that day's sale. If the number of newspapers stocked is too many then the unsold newspapers are wasted, likewise if the stock is low, the vendor loses opportunities to sell. The idea is similar in our setting whether there is an overage and an underage cost of going above or below q.

To model the setting as a newsvendor, we start by characterizing the excess demand. Define $V = D - S$ as a random variable denoting the excess demand beyond the renewable supply during the interval (t, k). So V is a continuous random variable which can be characterized using the distribution of D and S (that we can obtain using a distributional forecast). The cumulative distribution function (CDF) of V is $F_V(x) = \mathbb{P}\{V \leq x\}$ for any $x \in (-\infty, \infty)$ specifically noting it could be negative. For now assume that we have an expression for $F_V(x)$, but later we will provide an approach to computing it using the CDF of D and V.

If the excess demand is V and the microgrid commits a quantity q (which understandably could be negative denoting a selling to the utility grid), then the net cost per unit time is

$$dq + C^{RT} \max(V - q, 0) + B^{RT} \min(V - q, 0)$$

because for committing q there is a day-ahead price or credit of d, and there is either a real-time cost of C^{RT} to buy excess demand of $V - q$ or a real-time credit of B^{RT} to sell excess supply of $q - V$. Writing the expressions so that cost is positive and credit is negative, we get the expression for the net cost per unit time.

Let the expected value of the net cost per unit time be $\Psi(q)$. Our objective is to write down an expression for $\Psi(q)$ and minimize it to find the optimal value q^* (i.e., the minimum $\Psi(q)$ occurs at q^*). Taking expectations of the above net cost per unit time, we get the expected net cost per unit time $\Psi(q)$ as

$$\Psi(q) = \mathbb{E}[dq + C^{RT} \max(V - q, 0) + B^{RT} \min(V - q, 0)]$$

where the random variables are C^{RT}, B^{RT}, and V. Assuming C^{RT} and B^{RT} are independent of V, we get

$$\Psi(q) = dq + r\mathbb{E}[\max(V - q, 0)] + c\mathbb{E}[\min(V - q, 0)]$$

recalling that $\mathbb{E}[C^{RT}] = r$ and $\mathbb{E}[B^{RT}] = c$.

We rewrite the above expression for $\Psi(q)$ using the CDF of V as

$$\Psi(q) = dq + \int_{-\infty}^{\infty} [r \max(x - q, 0) + c \min(x - q, 0)] dF_V(x)$$

via the standard approach of taking expectation of a random variable using its CDF. To compute the integral, since the maximum and minimum functions can be defined differently before and after x taking the value of q, we have:

$$\begin{aligned}
\Psi(q) &= dq + \int_{-\infty}^{q} [r \max(x - q, 0) + c \min(x - q, 0)] dF_V(x) \\
&\quad + \int_{q}^{\infty} [r \max(x - q, 0) + c \min(x - q, 0)] dF_V(x) \\
&= dq + \int_{-\infty}^{q} [r \times 0 + c(x - q)] dF_V(x) + \int_{q}^{\infty} [r(x - q) + c \times 0] dF_V(x)
\end{aligned}$$

where the symbol "×" is a simple product. Continuing on by removing the zero terms we get

$$\begin{aligned}\Psi(q) &= dq+\int_{-\infty}^{q} cxdF_V(x)-\int_{-\infty}^{q} cqdF_V(x)+\int_{q}^{\infty} rxdF_V(x)-\int_{q}^{\infty} rqdF_V(x)\\ &= dq+c\int_{-\infty}^{q} xdF_V(x)-cq\int_{-\infty}^{q} dF_V(x)+r\int_{q}^{\infty} xdF_V(x)-rq\int_{q}^{\infty} dF_V(x)\\ &= dq+c\int_{-\infty}^{q} xdF_V(x)-cqF_V(q)+r\int_{q}^{\infty} xdF_V(x)-rq(1-F_V(q))\end{aligned}$$

and leaving the integrals as is for some of the cases. Since the maxima and minima occur where the derivative of $\Psi(q)$ is 0, we take the derivative and get $\Psi'(q)$ in terms of the probability density function $f(x)=F_V'(x)$ as

$$\begin{aligned}\Psi'(q) &= d+cqf(q)-cF_V(q)-cqf(q)-rqf(q)-r(1-F_V(q))+rqf(q)\\ &= d-cF_V(q)-r(1-F_V(q))\end{aligned}$$

since many of the terms cancel out.

Using $\Psi'(q^*)=(r-c)F_V(q^*)+d-r=0$, we get

$$F_V(q^*)=\frac{r-d}{r-c} \quad \text{or} \quad q^*=F_V^{-1}\left(\frac{r-d}{r-c}\right)$$

as the optimal values of q. We are not done yet as we need to ensure that it is the minimum. For that, since $\Psi''(q)=(r-c)f(q)\geq 0$ if $r\geq c$ for any q as the probability density function $f(q)$ is non-negative (which is a property of all probability density functions), q^* is at the minimum.

In summary, the optimal value of q that minimizes $\Psi(q)$ is q^* which satisfies

$$F_V(q^*)=\frac{r-d}{r-c} \quad \text{or} \quad q^*=F_V^{-1}\left(\frac{r-d}{r-c}\right)$$

which can be computed if we know the CDF of the excess demand. To illustrate that, next we present a numerical example and show how to compute the optimal value of q^*.

EXAMPLE TO ILLUSTRATE COMPUTING Q^*

Consider a microgrid that serves about 150 houses and has solar panels to satisfy electricity demand in the day time. The microgrid maintains time series data for the demand load in kW, solar power supplied in kW, the day-ahead cost of power (in \$ per kWh), the real-time cost of power (in \$ per kWh), the real-time credit for power (in \$ per kWh), the forecasted temperature, and the weather conditions across 15-minute intervals. A snapshot of such a dataset is provided in Table 5.1 (without the temperature and weather columns due to space). The table also uses the notation that have been presented thus far. Hence it would be useful to see numerical values of

Table 5.1
Snapshot of dataset to explain numerical data and notation

Index	datetime	(t,k)	D	S	$D-S$	C^{DA}	C^{RT}	B^{RT}
⋮	⋮	⋮	⋮	⋮	⋮	⋮	⋮	⋮
128	9/2/16 8:00am	(32, 1)	206.02	70.54	135.48	0.3492	0.4124	0.3172
129	9/2/16 8:15am	(32, 2)	199.26	75.21	124.05	0.3492	0.4531	0.2002
130	9/2/16 8:30am	(32, 3)	200.43	77.58	122.85	0.3492	0.4632	0.3175
131	9/2/16 8:45am	(32, 4)	189.23	79.49	109.74	0.3492	0.5219	0.2207
132	9/2/16 9:00am	(33, 1)	183.16	81.54	101.62	0.6120	0.5821	0.3102
133	9/2/16 9:15am	(33, 2)	179.77	73.79	105.98	0.6120	0.6724	0.2872
134	9/2/16 9:30am	(33, 3)	179.85	77.14	102.71	0.6120	0.6811	0.5921
135	9/2/16 9:45am	(33, 4)	190.45	52.06	138.39	0.6120	0.6114	0.5119
136	9/3/16 10:00am	(34, 1)	195.76	52.78	142.98	0.2142	0.4121	0.1191
⋮	⋮	⋮	⋮	⋮	⋮	⋮	⋮	⋮
232	9/4/16 10:00am	(58, 1)	209.37	92.41	116.96	0.1384	0.4287	0.0173
⋮	⋮	⋮	⋮	⋮	⋮	⋮	⋮	⋮
328	9/5/16 10:00am	(82, 1)	?	?	?	0.1693	?	?
⋮	⋮	⋮	⋮	⋮	⋮	⋮	⋮	⋮

the notations used in this chapter. The column (t,k) and the column $D-S$ have been computed subsequently (the original dataset did not have them).

It is worthwhile going over the data in Table 5.1 in some detail as it gives a clear perspective of the problem described thus far for the microgrid in question. The index column denotes the line number in the dataset starting with 0 that corresponds to 12:00am on 9/1/16. Note that the `datetime` column gives the date and time corresponding to the *start* of a 15-minute interval. The column (t,k) (defined in a previous section) is such that t is the hours elapsed since the start of the dataset and k corresponds to the quarter of the hour (hence takes values from 1 to 4). The column D and S correspond to the load demand and solar supply, both in kW. The column $D-S$ is simply the difference between the previous two columns. Note that at night time the S column would typically have zeros.

The day-ahead cost, real-time-cost, and day-ahead credit for each 15-minute interval are in the last three columns. Notice that the C^{DA} column values stay a constant throughout an hour, and that is because day-ahead prices are for an hour while real-time prices and credits are at 15-minute granularity in this problem setting (in other settings they could also be at 5-minute granularity). Notice that the realized real-time cost and credit can be higher or lower than the day-ahead price, but the expected value satisfies $r > d > c$. It is important to realize that the values in Table 5.1 are actual observed values and should not be confused with r, d, and c. We expand on that next.

The reason we have picked out index values of 232 and 328 is that we pretend that we are in the interval corresponding to index 232. So suppose that it is 10am on 9/4/16 right now. We know the current levels of demand load (D), solar supply (S), and all the costs as well as credits for that day. Our goal is to determine how many kW of electricity to commit for the next day for the 10–10:15am time slot. That day corresponds to the index value of 328. The reason there are "?" on certain columns of index 328 is because as of index 232, the values in index 328 are unknown. The only thing we know is the day-ahead price of 0.1693 (hence the name day-ahead price). This is also the d value for that row.

So for the index value of 328 with interval $(82,1)$ we wish to determine q^*, the optimal amount of power to commit from the utility grid. For that, we need the CDF of $D-S$ as well as the values r and c. For those, we respectively consider the time series of $D-S$, C^{RT}, and B^{RT} along with exogenous variables of temperature and weather predictions for the interval (82, 1). Getting r and c are relatively straightforward as they are simply the point forecast corresponding to C^{RT} and B^{RT} computed by minimizing the mean square error. For that we could use any of the forecasting approaches we considered in the previous chapter. Let us suppose that those forecasts are 0.4125 \$ per kWh and 0.0528 \$ per kWh, respectively.

Based on the above numerical values, we have

$$F_V(q^*) = \frac{r-d}{r-c} = \frac{0.4125-0.1693}{0.4125-0.0528} = 0.6761$$

where V is what we refer to as $D-S$. We will discuss forecasting the distribution of V and the concept of probabilistic forecasting in the next chapter. However, here we provide a rather simplistic view to see how we used the forecasting methods in the previous chapter to arrive at a crude distribution. We obtain a point forecast for $D-S$ using one of the forecasting approaches in the previous chapter by minimizing the root mean square error. Let us call this $\hat{\mu}_V$, as the estimate of the expected value of $D-S$ for that interval (82, 1).

As a crude estimate of the standard deviation of $D-S$ for that interval we simply use the root mean square error of the training set. Let us say that estimate is $\hat{\sigma}_V$. Let us suppose that $\hat{\mu}_V = 132.71$ and $\hat{\sigma}_V = 21.86$, both in units of kW. We assume that V is normally distributed with mean $\hat{\mu}_V$ and standard deviation $\hat{\sigma}_V$. Then we can invert the CDF relation $F_V(q^*) = 0.6761$ to get $q^* = 142.7$ kW using the Microsoft Excel command `=NORM.INV(0.6761, 132.71, 21.86)` as we are computing the inverse of the normal CDF at 0.6761 with mean and standard deviation of $\hat{\mu}_V$ and $\hat{\sigma}_V$ respectively.

Notice that we are committing from the utility grid an amount (142.7 kW) which is much higher than the expected value of power (132.71 kW). The reason is because the day-ahead price is much closer to the expected credit in real-time than expected price in real-time. So, we would be better off selling back to the utility grid, than buying more in real time. Thus we would prefer to commit to a higher value. However if the day ahead price was closer to the expected real-time price than the expected real-time credit, then we would have committed to an amount lesser than the mean.

Of course, all these rely on the accuracy of our estimates as well as choice of distribution.

GOING FROM q to Q

Recall that all the analysis done so far actually is for the interval (t,k). So the q^* we obtained is technically $q^*_{(t,k)}$ which is the optimal value of $q_{(t,k)}$. Having said that, what we need to commit to buying from the utility grid is $Q_{(t,k)}$ (with the understanding that if the number is negative it would mean committing to sell to the grid). As we stated before, the amount of power we commit would be a fixed constant during an hour, and so $Q_{(t,k)}$ is that amount of power to commit in the day-ahead market for hour t and it remains constant across k for a given t. But $q_{(t,k)}$ could vary by k for a given t. How do we go from $q^*_{(t,k)}$ to $Q^*_{(t,k)}$ where the latter is the optimal day-ahead commitment?

We take a straightforward approach of simply using the average value. So,

$$Q^*_{(t,k)} = \frac{1}{4}\sum_{i=1}^{4} q^*_{(t,i)}$$

for all $(t,k) \in \mathscr{A}_{test}$ where 4 is because there are four 15-minute intervals in an hour (it would have to modified to 12 if we are considering 5-minute intervals for real time). This averaging is indeed a crude approximation and is effective mainly because the values are not expected to be dramatically different in consecutive 15-minute intervals. However, if one were to consider a more granular approach, the following changes need to be made as explained next.

Let Q be the hourly commitment to buy/sell from the utility grid for hour t. We drop the t but keep the k in (t,k) in the remaining notation. Let d_k, r_k, c_k, and V_k correspond to the day-ahead cost, real-time cost, real-time credit, and the excess demand beyond solar supply, respectively in interval (t,k). Let $\Psi_k(Q)$ be the expected cost per unit time in interval (t,k) if we commit to Q kW of power during hour t. Then we have

$$\Psi_k(Q) = d_kQ + c_k\int_{-\infty}^{Q} x dF_{V_k}(x) - c_kQF_{V_k}(Q) + r_k\int_{Q}^{\infty} x dF_{V_k}(x) - r_kQ(1-F_{V_k}(Q))$$

for $k = 1,2,3,4$.

Suppose $\Psi(Q)$ is the expected net cost per hour for hour t, it is the sum

$$\Psi(Q) = \sum_{k=1}^{4} \Psi_k(Q)$$

for some Q which is the committed quantity of power during the hour t. Since $c_k \leq d_k \leq r_k$ for all k at all times t, we can obtain the minima as the value of Q where the derivative of $\Psi(Q)$ is 0. Taking that derivative

$$\Psi'(Q) = \sum_{k=1}^{4} \left[d_k - c_kF_{V_k}(Q) - r_k(1-F_{V_k}(Q))\right] \tag{5.1}$$

and setting it to zero, we can solve for Q^* numerically that satisfies $\Psi'(Q^*) = 0$. The expression is guaranteed to be at the minima for the same reason that we saw before when we solved for q^*. We do not present a numerical example here but we will do so in the next section.

CDF $F_V()$ USING CDF OF D AND S

The analysis predicted thus far is predicated upon knowing the CDF of the excess demand V in interval (t,k). The process of obtaining distributional forecasts of $V = D - S$ simply uses the time-series corresponding to the quantity $D - S$. However, considering that solar power is zero at night time and its patterns different from loads, there may be benefit in modeling D and S separately. This way, the model is conjectured to produce more accurate results. However, the approach of going from distributions of D and S to that of $D - S$ is a bit involved with significant use of probabilistic modeling and analysis (and it is unclear if there is a significant benefit in doing so). The reader is welcome to skip the rest of this section and return to it later, if necessary.

To characterize the excess demand random variable $V = D - S$ with CDF $F_V(x)$ that we use to obtain the optimal quantity of power to commit a day ahead, we can start with the distributions of D and S. In other words, can we obtain $F_V(x)$ using the CDFs of D and S. Suppose we know the joint CDF $F_{D,S}(y,z) = \mathbb{P}\{D \leq y, S \leq z\}$ for all $y \geq 0$ and $z \geq 0$ as D and S are non-negative. It is entirely possible that D and S are correlated (positively or negatively) in a given interval (t,k). So, if we know $F_{D,S}(y,z)$, then we can obtain the following:

(a) The marginal CDF $F_D(y)$, mean $\mu_d = \mathbb{E}[D]$, and variance $\sigma_d^2 = \mathbb{V}[D]$;
(b) The marginal CDF $F_S(z)$, mean $\mu_s = \mathbb{E}[S]$, and variance $\sigma_s^2 = \mathbb{V}[S]$;
(c) The correlation coefficient

$$\rho = \frac{\mathbb{E}[DS] - \mathbb{E}[D]\mathbb{E}[S]}{\sigma_d \sigma_s}$$

such that $\rho \in [-1,1]$.

The above is a probabilistic approach, however, in a data-driven approach we first obtain estimates of μ_d, σ_d^2, μ_s, σ_s^2, and ρ directly from the training data. Then we obtain the mean and variance of the excess demand random variable V. For that we use estimates of μ_d, σ_d^2, μ_s, σ_s^2, and ρ we wish to obtain the mean and variance of V. Since $V = D - S$, we have $\mathbb{E}[V] = \mathbb{E}[D - S] = \mu_d - \mu_s$. Also, we get the variance as follows:

$$\begin{aligned}
\mathbb{V}[V] &= \mathbb{V}[D-S] = \mathbb{E}[(D-S)^2] - (\mu_d - \mu_s)^2 \\
&= \mathbb{E}[D^2] + \mathbb{E}[S^2] - 2\mathbb{E}[DS] - (\mu_d - \mu_s)^2 \\
&= \sigma_d^2 + \mu_d^2 + \sigma_s^2 + \mu_s^2 - 2\mathbb{E}[DS] - \mu_d^2 - \mu_s^2 + 2\mu_d\mu_s \\
&= \sigma_d^2 + \sigma_s^2 - 2\rho\sigma_d\sigma_s
\end{aligned}$$

which is always guaranteed to be non-negative since $\rho \in [-1,1]$.

As a simple second-order approximation, we could just assume that V is normal with the above mean and variance. Note that when D and S are independent, $\rho = 0$. In fact, the above results are exact if D and S are Normally distributed. Otherwise, if D and S are according to some distribution with the aforementioned joint CDF, then we can obtain the CDF of the excess demand random variable as follows. Say we wish to compute $F_V(x_0) = \mathbb{P}\{V \leq x_0\}$ for an arbitrary x_0. We derive an expression for the case $x_0 > 0$ with the understanding that a similar result can be derived for $x_0 < 0$.

Since $V = D - S$, we have for any $x_0 > 0$

$$\begin{aligned}
\mathbb{P}\{V \leq x_0\} &= \mathbb{P}\{D - S \leq x_0\} = \int_0^\infty \mathbb{P}\{D - S \leq x_0 | S = z\} dF_S(z) \\
&= \int_0^\infty F_D(x_0 + z | S = z) dF_S(z) = \int_0^\infty \int_0^{x_0+z} f_{D|S}(y|z) dy dF_S(z) \\
&= \int_0^\infty \int_0^{x_0+z} f_{D,S}(y,z) dy dz
\end{aligned}$$

where $f_{D,S}(y,z)$ is the joint probability density function of D and S which can be obtained by taking the partial derivative of the joint CDF $F_{D,S}(y,z)$ with respect to y and z. When D and S are *independent*, we get

$$\mathbb{P}\{V \leq x_0\} = \int_0^\infty F_D(x_0 + z) dF_S(z)$$

although they being independent is unlikely.

5.3 SYSTEM WITH STORAGE: TWO-STAGE OPTIMIZATION

In this section we consider a microgrid with an energy storage device (ESD) capable of storing up to $\mathscr{K}$ kWh of energy. Other than that, the setting and notation of the previous section continues here. Here too we need to make a day-ahead hourly commitment for power from the utility grid for each hour of the day. However, we make a small adjustment here. Every day at 4pm the day-ahead prices and credits are announced. However, it is announced for each of the 24 hours of the *next* day, i.e., from hour 0 to hour 23. So, the forecasts are day-ahead but not 24 hours ahead. The forecast length varies from hour to hour and some hours are larger than 24 hours ahead. In that sense this is another difference, besides having an ESD.

The real-time prices, credits, demand, and solar supply will be revealed at the beginning of each of the 15-minute intervals of the next day. However, at the day-ahead 4pm time, we will use point or interval estimates available at that time. At that 4pm time we know the day-ahead cost to commit power, so we know $C^{DA}_{(t,k)}$ for all $t = 0, \ldots, 23$ (assuming there is just a single test day, which is the case here) and $k = 1,2,3,4$ with the understanding that $C^{DA}_{(t,k)}$ remains fixed under all k for a given t. Using that information we first need to make the decision regarding how much power to commit at 4pm for the next day. We also need to determine during each of the 15 minute interval how much power to buy or sell at real time prices.

To make day-ahead decisions we need to forecast demand, supply, and real-time costs as well as credits. In the previous section where there was no ESD, the real-time decisions were straightforward as in real time we simply buy or sell power at the real-time cost or credit value to exactly satisfy our demand in the 15-minute interval. When there is an ESD, we may wish to buy more and store or sell more and deplete the ESD. So we would need to formulate and solve an optimization problem. Although the day-ahead and real-time problems are intertwined, we will present them separately for ease of exposition. We will revisit the problem description after going over some of the modeling nuances that need to be considered.

The only thing that prevents us from solving independent problems day by day is the amount of power stored in the ESD that carries over. We circumvent the situation of using more than 1 day's horizon due to the ESD level linkage by requiring the ESD level to be at least as much as the previous day in the day-ahead problem and imposing a large penalty cost if the next day's energy storage level is smaller than the current day's level. We will explain this while describing the two problems. This is more of an issue in the real-time decisions as not imposing a penalty would make the methods greedily use up all the ESD charge when the horizon is only one day.

The linking constraint, while imposes complications, is not the only issue to be concerned about while having a system with an ESD. The ESD has many other nuances that require a few more notation. The ESD is not 100% efficient, and so the inefficiencies in charging (denoted as η_{char} which is less than 1) and discharging (denoted as η_{dis}, also less than 1) need to be incorporated. That means if we input z kWh to the battery, the storage level increases by $z\eta_{char}$. Likewise while discharging, to get an output of z kWh the ESD level would go down by z/η_{dis}. However, this inefficiency has been steadily improving in newer technologies.

In addition, the ESD system has a charging capacity c_{char} kW and discharging capacity c_{dis} kW. Many commercial batteries have a charge and discharge rates available for the battery. But more often than not, the bottleneck in charging is not the battery capacity but those of the inverter. Inverters are power electronics that convert DC from battery to AC for the grid (discharging) as well as convert AC from grid to DC to store in battery (charging). These inverters limit the maximum power one can push in/out of a battery, regardless of battery capacity. The number of hours to change a battery from half full to full is $0.5\mathscr{K}/c_{char}$ hours.

Using all this information, our goal is to solve an optimization problem that would tell us how much to commit each hour of a day from the grid a day in advance (i.e., 4pm the previous day when day-ahead prices are revealed), as well as how much of the real-time excess demand (beyond the committed amount) to satisfy from the utility grid and the ESD. Hence the title of this section has the words "two-stage optimization". Objective is to minimize expected cost subject to satisfying constraints for flow balance, ESD energy level adjustment, charging and discharging capacities, and any restrictions on the decision variables.

There are multiple ways to solve this problem. One option that is well documented in the literature including some of the author's collaborative work uses two-stage stochastic program with here-and-now day ahead decisions, and second-stage decisions corresponding to real-time. However, the concepts in stochastic programming are too vast to cover in this book. Hence we take a different approach here and refer the reader to various research articles to solve this problem. We take a rather simple approach by solving a separate day-ahead problem and determine the commitments, and then use those commitments in making the real-time decisions sequentially at the beginning of each 15-minute intervals.

DAY-AHEAD COMMITMENT: EXTENDING THE NEWSVENDOR

At 4pm on a particular day, given the day-ahead prices for each hour of the next day, the objective is to determine the amount of power to commit to or from the utility grid. Suppose we have a training data of several days with hour and 15-minute interval represented as (t,k) in the set $\mathscr{A}_{train}$. Likewise, there is a test data of 1 day, so $\mathscr{A}_{test} = \{(0,1),(0,2),(0,3),\ldots,(23,3),(23,4)\}$. For all $(t,k) \in \mathscr{A}_{test}$, we are given the day-ahead cost $C^{DA}_{(t,k)}$ which is not a function of k as the cost or reward is hourly.

Using the training data up until 4pm of the day before the test data, we code up an exponential smoothing model to forecast the net load as the difference between the demand and solar supply $D_{(t,k)} - S_{(t,k)}$ in terms of its mean $\mu^{D-S}_{(t,k)}$ and standard deviation $\sigma^{D-S}_{(t,k)}$ and denote them as $\hat{\mu}^{D-S}_{(t,k)}$ and $\hat{\sigma}^{D-S}_{(t,k)}$. The reason to use the net load and not the individual demand and supply is because we only need the net load, but most importantly, the supply values go to zero at night time and thus an exponential smoothing model is not conducive.

Now we briefly explain how we code up an exponential smoothing model to obtain the estimates of $\hat{\mu}^{D-S}_{(t,k)}$ and $\hat{\sigma}^{D-S}_{(t,k)}$. For that, refer to Cell 4 shown in Figure 3.8 where we compute intervals for forecasts made for $h = 16$ as the horizon. We need to make suitable changes here. In line 2, the time series is now that corresponding to the net load $D_{(t,k)} - S_{(t,k)}$ for all $(t,k) \in \mathscr{A}_{train}$ by being careful in stopping the time series at 4pm. Since we have 24 hours of data (as opposed to 8 hours in the solar forecasting data in Figure 3.8), we use 96 as the number of seasonal periods in line 5 (this must be incorporated in the code).

Then in line 8 we use $h = 128$. There are two important things to note. First is that $h = 128$ because at 4pm we need to make predictions for ninety-six 15-minute intervals the next day plus 32 intervals for the 8 hours remaining in the day after 4pm. The second thing is we use predictions from the 33rd to the 128th prediction, not the entire horizon. If we have training data for n whole days, since we need to stop at 4pm, in line 11 we would use $T = 96(n-1)+64$ as the first $n-1$ days would use all 96 data points while the last day would only use until 4pm which is 64 data points. Also we simply choose $i = 0$ in line 17. There is no change in line 20.

In line 23, one could try various models. However, in the results we present, we did not consider trend and the seasonality was modeled as additive. In practice, one could try all combinations and use the one that produces the best results. The remaining lines are the same except the last two. Here since we are not interested in an interval forecast but simply the estimated standard deviation, we can drop lines 29 and 30. Instead we could just say `std_est = simulation_results.std(axis = 1)`. To obtain $\hat{\mu}^{D-S}_{(t,k)}$ and $\hat{\sigma}^{D-S}_{(t,k)}$ for all $(t,k) \in \mathscr{A}_{test}$, one should be careful in extracting the last 96 elements of `forecasts` and `std_est` respectively.

The real time cost $C^{RT}_{(t,k)}$, and the real-time credit $B^{RT}_{(t,k)}$ can also be similarly fitted using the training data and predicted for the testing data. Note that we only require the mean estimates and hence we do not need to simulate. We can simply stop at line 27 in Figure 3.8. In fact, it would be useful to just write one python notebook cell for all the predictions. For the test data we can use $\hat{C}^{RT}_{(t,k)}$ and $\hat{B}^{RT}_{(t,k)}$ for the forecasted cost and credit, respectively in real time for all $(t,k) \in \mathscr{A}_{test}$. We use the mean and standard deviation forecasts of the net load, the forecasts of the real time costs and credits, as well as the day ahead costs to determine the day-ahead commitment.

For that day-ahead commitment, we use the results from Section 5.2 and combine with the effect of adding an ESD. When there is no storage, for any $(t,k) \in \mathscr{A}_{test}$, we can compute $q^*_{(t,k)}$ as the solution to

$$F_V(q^*_{(t,k)}) = \frac{\hat{C}^{RT}_{(t,k)} - C^{DA}_{(t,k)}}{\hat{C}^{RT}_{(t,k)} - \hat{B}^{RT}_{(t,k)}} \tag{5.2}$$

where $F_V(\cdot)$ is the CDF of a normal random variable with mean $\hat{\mu}^{D-S}_{(t,k)}$ and standard deviation $\hat{\sigma}^{D-S}_{(t,k)}$ with the understanding that V denotes $D-S$ as stated in Section 5.2.

While the above is a good estimate of how much we could commit a day in advance, the question is if we can do better when there is storage. The key idea is to buy a little more when the day-ahead price is low and sell that power when the price become high. This way when the price is high, we can tap into the ESD. Then our final hourly commitment would be the sum of what we would commit to when there is no storage plus the adjustment due to storage. Next we present the analysis to make that adjustment. For that we consider a requirement that if we started the day with u_{init}, then we must end the day also with u_{init} amount of charge in the ESD. This way it forces one to not greedily use up the ESD.

We introduce some notation. Since this is only at the hourly level, we only need t to denote the hour and simply use $k = 1$ (as the values are identical throughout the hour), for the day-ahead cost $C^{DA}_{(t,k)}$. We define two non-negative decision-variables W^c_t and W^d_t denoting the amount of power used to charge and the amount of power discharged from the ESD during hour t. Also, let u_t be an auxiliary variable that denotes the ESD charge level at the beginning of hour t. All other variables η_{dis}, η_{char}, $\mathscr{K}$, c_{dis}, and c_{char} are ESD-related quantities defined earlier.

We formulate a linear program to solve for W^c_t and W^d_t which informs us how much to commit during each hour. We will discuss the commitments later but first let us look at the problem we wish to solve to obtain W_t^c and W_t^d:

$$\text{Minimize } \frac{1}{24}\sum_{t=0}^{23} C^{DA}_{(t,1)}(W_t^c - W_t^d) \quad (5.3)$$

Subject to

$$u_0 = u_{init} \quad (5.4)$$

$$u_{24} = u_{init} \quad (5.5)$$

$$u_{t+1} = u_t - W_t^d/\eta_{dis} + W_t^c\eta_{char} \quad \forall t \in \{0,1,2,\dots,23\} \quad (5.6)$$

$$0 \le W_t^d \le c_{dis} \quad \forall t \in \{0,1,2,\dots,23\} \quad (5.7)$$

$$0 \le W_t^c \le c_{char} \quad \forall t \in \{0,1,2,\dots,23\} \quad (5.8)$$

$$0 \le u_t \le \mathscr{K} \quad \forall t \in \{0,1,2,\dots,23\} \quad (5.9)$$

which is explained in detail below.

Since W_t^c and W_t^d denote the power drawn from the utility grid to charge the ESD and sold to the utility grid after discharging the ESD, the quantity $W_t^c - W_t^d$ is the net purchase from the grid during hour t. Our objective is to minimize the average net purchase cost over a day and hence we multiply $W_t^c - W_t^d$ by $C^{DA}_{(t,1)}$ and average over all t in a day in Equation (5.3). We use $C^{DA}_{(t,1)}$ instead of $C^{DA}_{(t,k)}$ as for any k and given t, $C^{DA}_{(t,k)}$ remains a constant and so we select $k = 1$. Notice that $W_t^c - W_t^d$ could be both positive or negative, however there are some constraints imposed that regulates the values $W_t^c - W_t^d$ can take.

Equation (5.4) states that the variable u_0 corresponding to the ESD level at the beginning of hour 0 must be equal to u_{init}, the predicted initial level. Likewise, Equation (5.5) states that at the end of the day, the ESD level must be what it was at the beginning of the day. As an alternate formulation, one could change the $=$ to a $\ge$ but would not make a difference unless some of the day-ahead costs are negative. Then, Equation (5.6) essentially performs the energy balance between the beginning of an hour to the next hour. At the beginning of hour t, we start with u_t kWh of energy in the ESD. Then if we sell W_t^d to the grid, then we would discharge the ESD by W_t^d/η_{dis} to account for inefficiency.

Likewise, if we were to purchase W_t^c amount of power during hour t, we would charge the ESD by $W_t^c\eta_{char}$. Thus u_{t+1} would be u_t minus the energy that is used to discharge plus the energy that is obtained while charging. It is important to note that the units of W_t^c and W_t^d are kW and the η_{dis} and η_{char} are dimensionless. However, what is missing is the 1 hour that is multiplied during which the ESD gets charged or discharged. So, when we consider other lengths of time, there would be an explicit multiplier of number of hours, and thereby the units would match. One would see this if we use a similar expression for the real-time energy balance.

The inequality (5.7) essentially states that W_t^d needs to be non-negative but the amount discharged can be no more than c_{dis}. Likewise, the inequality (5.8) states that W_t^c needs to be non-negative but the amount charged can be no more than c_{char}. Finally, inequality (5.9) states that at the beginning of any hour, the ESD level needs to be between its minimum of 0 and its maximum of $\mathscr{K}$. Also, the expression $\forall t \in \{0,1,2,\dots,23\}$ in the last four expressions is to say that all those expressions hold for all t. In other words, instead of writing 96 lines, we write 4.

```
1 import cvxpy as cp
2
3 def solve_battery_lp(prices, eta_c, eta_d, kappa_cap, c_dis, c_char, u0):
4     T = len(prices)
5
6     p_c = cp.Variable(T, nonneg=True)  # charge power
7     p_d = cp.Variable(T, nonneg=True)  # discharge power
8     U = cp.Variable(T+1)               # state of charge
9
10    constraints = []
11
12    # Initial state of charge (SoC)
13    constraints.append(U[0] == u0)
14
15    # SoC dynamics
16    for t in range(T):
17        constraints.append(U[t+1] == U[t] + eta_c * p_c[t] - (1 / eta_d) * p_d[t])
18
19    # SoC bounds
20    constraints += [U >= 0, U <= kappa_cap]
21
22    # Power bounds
23    constraints += [p_c <= c_char, p_d <= c_dis]
24
25    constraints += [U[24] == u0] # End of day charge level = start
26
27    # Objective: minimize cost = price @ net discharge
28    net_power = p_c - p_d
29    objective = cp.Minimize(prices @ net_power)
30
31    prob = cp.Problem(objective, constraints)
32    prob.solve()
33
34    if prob.status not in ["optimal", "optimal_inaccurate"]:
35        raise ValueError(f"Problem status: {prob.status}")
36
37    return net_power.value
38
39 V_opt = solve_battery_lp(prices, eta_c=0.9, eta_d=0.9, kappa_cap=200,
40                          c_dis=25, c_char=25, u0=100)
```

Figure 5.2 Python notebook cell for optimal power to purchase from grid for ESD

Several mathematical programming packages solve such linear programs and they can be solved extremely fast. Even in python there are various libraries that can be used. We use a convex programming library here as a linear program is a special case of a convex program. An example of that is in Figure 5.2 to solve the problem at hand. The first line in that cell is to import the library `cvxpy`, it may require a pip install. Lines 3 to 37 provide a function into which if we provide the day-ahead costs as a vector of 24 values (`prices`), η_{char}, η_{dis}, $\mathscr{K}$ (called `kappa_cap`), c_{dis}, c_{char}, and u_{init} (as `u0`), the function would return a vector of optimum values of $W_t^c - W_t^d$ for all t from 0 to 23.

An example of using the function is in lines 39 and 40 of the cell shown in Figure 5.2 where numerical values of all inputs (except the day-ahead price vector) are shown. Then the resulting vector is mapped to `V_opt`. The function itself is simply describing the formulation. In line 4 we start with defining T which equals 24 (the length of the vector called `prices`). In lines 6, 7, and 8 the three variables are stated (`p_c` is W_t^c vector for all t, likewise `p_d` is the vector of W_t^d, and U the vector of u_t with the understanding that u has 25 values from 0 to 24). Line 10 starts with an empty constraint set. Then we append to the constraints. Lines 13, 17, 20, 25 correspond respectively to (5.4), (5.6), (5.9), and (5.5).

Line 21 has the right-hand-sides of both (5.7) and (5.8), and the left-hand-side is stated in lines 6 and 7 as a non-negative (`nonneg=True`) variable. Having set up the constraints, the next it to define the objective. For that the product of the day-ahead prices and the net power is used in line 29 and they directly correspond to (5.3). In line 31 we state that the convex optimization problem is stated with the objective and the constraints. Then in line 32, the problem is asked to be solved. In line 32 for larger problems, one can also add the solver name, especially something that is specially designed for linear programs. Line 34 essentially states that if there is no solution, then a flag be raised.

Suppose the resulting optimal value of solving the linear program is $W_t^{*c} - W_t^{*d}$, i.e., the value that minimizes the objective and also satisfies the constraints. This tells us how much more (if positive) or less (if negative) we should get from the utility grid at time t than what we would have committed a day ahead, if we did not have a storage (i.e., ESD). Hence, we immediately write down the **day-ahead commitment** of grid power W_t^{dac} for time t as

$$W_t^{dac} = W_t^{*c} - W_t^{*d} + \frac{1}{4}\sum_{k=1}^{4} q^*_{(t,k)}$$

where $q^*_{(t,k)}$ is the solution to Equation (5.2) corresponding to the day-ahead commitment if we did not have a storage. As an alternative, one could use the approach in Equation (5.1).

Simulated Example

To explain the results and verify that it conforms to our intuition, we consider a dataset that was simulated with relatively low noise. Forty-five days of data was generated starting September 1, 2023 and ending October 15, 2023. For the demand we considered two peaks each day, one in the morning and one in the evening and added some random noise. For the supply we created a sinusoidal pattern with noise during the day time to depict solar irradiance and added low amount of noise to indicate clouds. Likewise day ahead prices were generated with some diurnal patterns and real-time costs as well as credits enveloped the day-ahead prices with some light amount of noise. All data were generated in 15-minute intervals across the whole day. Also, we immediately computed the net load as the difference between demand and supply.

Using this data we performed the analysis considering the first 35 days (stopping at 4pm on the 35th day) as training and predicted using exponential smoothing the

(a) Real time costs (b) Real time credits

Figure 5.3 Exponential Smoothing Prediction versus Ground Truth for last 2 Days of Dataset

net load, real-time cost, and real-time credits in 15-minute intervals for the 36th day. After the 36th day's predictions are made, the training window moves by 1 day and we make predictions for the 37th day. This process continues and we perform 10 tests on the last 10 days of the data. We present the results using figures, as opposed to accuracy metrics. Figure 5.3 shows the last two days' test data with ground truth of real time cost and real time credit plotted against exponential smoothing forecasts. Since the noise is very low, the forecasts are close to the ground truth values. The results for net load (difference of demand and supply) are also similar but not presented here.

For the day-ahead commitment at 4pm of the previous day, following our approach outlined above, we first compute $q^*_{(t,k)}$ using Equation (5.2) as we have forecasts for the real-time costs $\hat{C}^{RT}_{(t,k)}$, real-time credits $\hat{B}^{RT}_{(t,k)}$, mean net load $\hat{\mu}^{D-S}_{(t,k)}$, and standard deviation of net load $\hat{\sigma}^{D-S}_{(t,k)}$, as well as the day-ahead costs $C^{DA}_{(t,k)}$. For each hour t we denote Q^*_t as the average of $q^*_{(t,k)}$ over all k. Figure 5.4 (a) depicts $\hat{C}^{RT}_{(t,k)}$, $\hat{B}^{RT}_{(t,k)}$, and $C^{DA}_{(t,k)}$ over all t and k across 1 day of testing data. Notice that for each t and k, the condition $\hat{C}^{RT}_{(t,k)} \geq C^{DA}_{(t,k)} \geq \hat{B}^{RT}_{(t,k)}$ is satisfied.

Also notice from Figure 5.4 (a) that $C^{DA}_{(t,k)}$ remains constant over an hour while the other two forecasted real-time costs vary within an hour, as they should. In Figure 5.4 (b) a graph of the forecasted average net load (demand minus solar supply) $\hat{\mu}^{D-S}_{(t,k)}$ and Q^*_t (the average of $q^*_{(t,k)}$ over all k using Equation (5.2)) are shown. Recall that the result requires that the real-time cost and credit each be independent of the net load. Upon computing the Pearson correlation coefficient, both were less than 5% indicating that the values are nearly independent. Notice that Q^*_t (which is the amount of power that would be committed from the utility grid if we had no storage) almost directly overlaps with the expected load $\hat{\mu}^{D-S}_{(t,k)}$.

While it is anticipated that Q^*_t mimics $\hat{\mu}^{D-S}_{(t,k)}$, direct overlap may not necessarily happen. The reason there is direct overlap here is that for all t, the difference $\hat{C}^{RT}_{(t,k)} - C^{DA}_{(t,k)}$ is nearly equal to $C^{DA}_{(t,k)} - \hat{B}^{RT}_{(t,k)}$ as evident in Figure 5.4 (a). When that happens the right-hand-side of Equation (5.2) is half and then $q^*_{(t,k)}$ equals to the mean of the normal distribution which indeed is $\hat{\mu}^{D-S}_{(t,k)}$. When they are not equal and the values deviate as well as how large of small the standard deviation estimate $\hat{\sigma}^{D-S}_{(t,k)}$ is, the Q^*_t could go above or below the $\hat{\mu}^{D-S}_{(t,k)}$ curve.

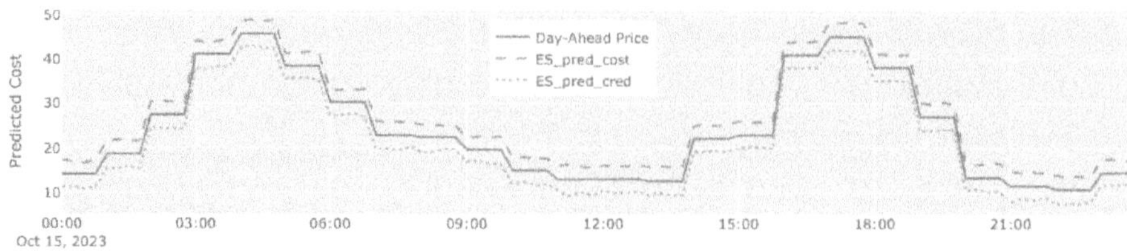

(a) Day-ahead costs $C^{DA}_{(t,k)}$ and forecasts of real time costs $\hat{C}^{RT}_{(t,k)}$ and credits $\hat{B}^{RT}_{(t,k)}$ for all t and k in a day

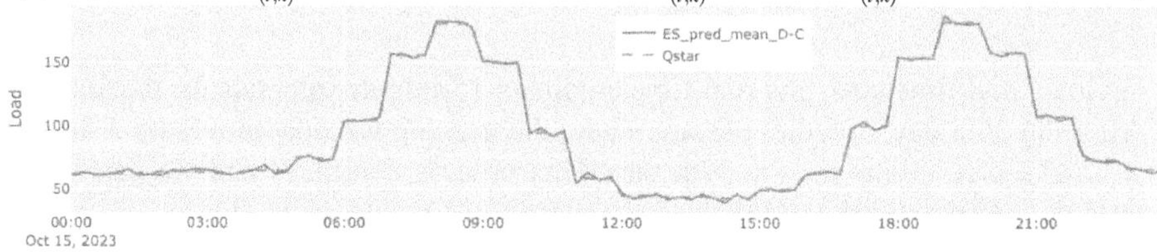

(b) Forecast of average net load $\hat{\mu}^{D-S}_{(t,k)}$ and commitment under no storage Q^*_t for all t and k in a day

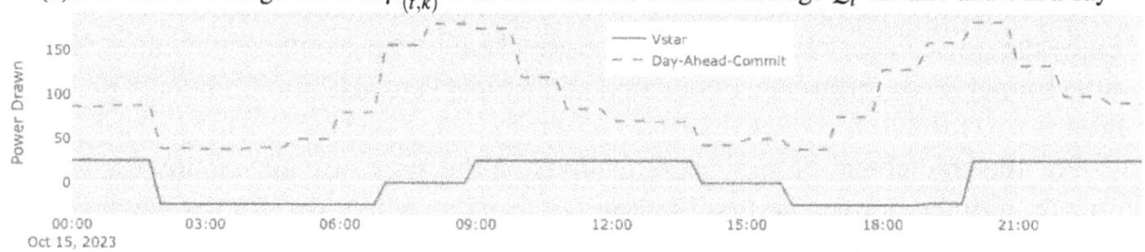

(c) $W^{*c}_t - W^{*d}_t$ and day-ahead commitment W^{dac}_t with storage for all t and k in a day

Figure 5.4 Inputs, intermediate calculations, and outputs for hourly day ahead commitment

Next we go over Figure 5.4 (c) where $W^{*c}_t - W^{*d}_t$ (called `Vstar` in the legend) and the resulting day-ahead commitment are presented for all t in a day. The numerical values used are those in Figure 5.2 except for u_{init} which was sampled from a normal distribution with mean 100 and standard deviation 20. This would result in each of the 10 test days to start with a different initial level of ESD. Notice that $W^{*c}_t - W^{*d}_t$ predominantly takes on 3 values: 25, 0, and -25 which corresponds to $W^{*c}_t - W^{*d}_t$ equalling c_{dis}, neither charging nor discharging, and $-c_{char}$. This is intuitive as when u_t is neither full nor empty, and if we choose to charge or discharge in an hour, we might as well do that at the fullest extent.

Also, recall from the linear programming formulation, $W^{*c}_t - W^{*d}_t$ only depends on the day-ahead costs and not the other costs, credits, or net load. This is reflected in Figure 5.4 (c) by comparing against the day-ahead price in Figure 5.4 (a). However, the dependence is in an opposite sense. This should not be surprising, considering the objective function. However, from an intuitive standpoint, note that when the price is high we discharge, when it is low we buy and charge, and we do nothing in medium values. Then, upon adding $W^{*c}_t - W^{*d}_t$ to Q^*_t, we get the *day-ahead commitment* W^{dac}_t which is also shown in Figure 5.4 (c). With this day ahead commitment, next we will see how one would go about course-correcting in real time.

REAL-TIME DECISIONS: USING REINFORCEMENT LEARNING

In this section, we are given the day ahead commitment W_t^{dac} for each hour of t each day using the analysis presented in the day-ahead decisions description above. Here we decide for each 15-minute interval $(t,k) \in \mathscr{A}_{test}$ how much power to buy from or sell to the utility grid $X_{(t,k)}$ (if positive, it denotes buy; else, it denotes sell), how much power to draw from the solar PV cells $Y_{(t,k)}$, and how much power to draw from or deposit to the ESD, $Z_{(t,k)}$ (if positive, denotes discharge; else, it denotes charge). As described in the remark in Section 5.1, the objective is to determine $X_{(t,k)}$, $Y_{(t,k)}$, and $Z_{(t,k)}$ for all $(t,k) \in \mathscr{A}_{test}$ so that the average cost per hour in the entire test set is minimized.

At the beginning of each 15-minute interval we assume that we know the demand, solar supply, real-time cost, and real-time credit for that interval. However, for future intervals those are not known with certainty. We also know the day-ahead commitment for that interval, the ESD charge level at the beginning of the 15-minute interval, and all the parameters defined earlier with respect to the ESD. Thus, for any $(t,k) \in \mathscr{A}_{test}$, given demand $D_{(t,k)}$, solar PV supply $S_{(t,k)}$, real-time cost $C^{RT}_{(t,k)}$, real-time credit $B^{RT}_{(t,k)}$, ESD level at the beginning of the interval $U_{(t,k)}$, and day-ahead commitment W_t^{dac}, we wish to obtain the optimal values of $X_{(t,k)}$, $Y_{(t,k)}$, and $Z_{(t,k)}$.

The state of the system at the beginning of interval (t,k) can be characterized by a 4-dimensional vector (or a 4-tuple) corresponding to the net real-time load $D_{(t,k)} - S_{(t,k)} - W_t^{dac}$ as the first dimension and the other three dimensions being $C^{RT}_{(t,k)}$, $B^{RT}_{(t,k)}$, and $U_{(t,k)}$. When this state is revealed at the beginning of interval (t,k), we need to take an action $Z_{(t,k)}$ which is to decide how much to discharge (positive-valued $Z_{(t,k)}$) or charge (negative-valued $Z_{(t,k)}$). Although we have three decisions to make, our action space is only one of them. We explain the reasoning next.

The one variable action is because $Y_{(t,k)} = S_{(t,k)}$ as it would be possible to sell any amount of excess power to the utility grid, if $B^{RT}_{(t,k)} \geq 0$ there is no need to curtail and all the solar PV power can be used or sold to the utility grid. Also, once $Z_{(t,k)}$ is known, we can simply compute $X_{(t,k)} = D_{(t,k)} - S_{(t,k)} - W_t^{dac} - Z_{(t,k)}$ the expression ensures energy balance during the 15-minute interval since all the load is satisfied by solar PV power, day-ahead commitment, ESD, and real-time power purchased from the utility grid. It is important to understand that all numbers other than the demand and supply, can be positive or negative.

Motivated by the results for the simulated example in the previous subsection, here we consider only three actions, i.e., based on the state at the beginning interval (t,k), the actions which are possible values of $Z_{(t,k)}$ are only 3. In particular, we consider discharge as much as possible ($Z_{(t,k)} = \min(4\eta_{dis}U_{(t,k)}, c_{dis})$), charge as much as possible ($Z_{(t,k)} = -\min(4(\mathscr{K} - U_{(t,k)})/\eta_{char}, c_{char})$), or do nothing ($Z_{(t,k)} = 0$). The reason there is a "4" pre-multiplying $U_{(t,k)}$ is because $U_{(t,k)}$ is in kWh and the intervals are $1/4$ hours.

Also, in case of discharge as much as possible is that typically the discharge rate is c_{dis} kW. In a 15-minute interval while discharging, $U_{(t,k)}$ can reduce at the most by $c_{dis}/(4\eta_{dis})$. The charging as much as possible case works in a similar fashion with the maximum kWh that can be gained by the ESD is $\mathscr{K} - U_{(t,k)}$. One can update the ESD level for the next state as $U_{(t,k)} - Z_{(t,k)}/(4\eta_{dis})$ if the action was to discharge as much as possible; as $U_{(t,k)} - Z_{(t,k)}\eta_{char}/4$ if the action was to charge as much as possible; as $U_{(t,k)}$ if the action is to do nothing.

Recall that we can write down $X_{(t,k)} = D_{(t,k)} - S_{(t,k)} - W_t^{dac} - Z_{(t,k)}$ and the instantaneous reward in taking action $Z_{(t,k)}$ is the amount sold to the grid (with negative meaning cost instead of reward). If $X_{(t,k)} \leq 0$ implying net sale of power, the instantaneous reward is -$X_{(t,k)}B^{RT}_{(t,k)}$. However, if $X_{(t,k)} > 0$ implying net buying of power, the instantaneous reward is -$X_{(t,k)}C^{RT}_{(t,k)}$. That said, the decision to be made is what action to select when a state is revealed at the beginning of an interval. Since we wish to maximize the average reward across the entire test data, it would not be smart to greedily select the action that maximizes the instantaneous reward. Instead one would have to select the optimal actions with the long-term benefits in mind.

There are a few ways to obtain good actions to take in each of the states. One of them is the concept of *reinforcement learning* (RL) which is a type of machine learning technique. In RL, an agent learns the optimal actions to take in an environment by initially trying out various actions randomly and obtaining rewards or penalties for them. The idea is to maximize the cumulative reward over the entire horizon of time. The main components of RL are states, actions, and rewards. Although there are many RL methods, in this book we use Q-learning which is arguably one of the simplest of RL methods. It uses a value-based method through a so called Q-table to obtain the action-selection policy.

A Q-table has rows corresponding to the states and columns corresponding to actions. Each element of the Q-table has an estimate of the expected future reward for taking the action corresponding to the column of the element when the state corresponds to the row of Q-table element. To build such a Q-table, it is necessary that the states and action are discrete and finite. Recall that we have a 4-dimensional (or 4-tuple) state representation. We would need to discretize each dimension. We consider 10 buckets for each of the 4 dimensions. Here we arbitrarily choose 10, but in an ideal circumstance one would tune this like a hyper-parameter. Also, the buckets are determined using historical data.

If we divide each dimension of our state into 10 buckets, we would have up to a total of 10,000 rows in the Q-table. Also, we would have exactly 3 columns corresponding to the 3 actions of discharge as much as possible, charge as much as possible, or do nothing. Although the Q-table is designed for all possible states, it only stores the states that have been encountered. For example, it is unlikely to have a state with a high real-time credit and low real-time cost, so that state would never be encountered. In theory it is possible to have all states but it is far more efficient to only list those that were encountered.

Next we discuss how we would populate the Q-table with an estimate of expected future reward for a particular state (row) and action (column). For this we need a training data set. For all states we initialize the estimate as an arbitrary value (in our case we use 0 as the initial entry for all actions in a state that has not been visited). Then the algorithm proceeds by determining the state and dropping it into the appropriate discretized bucket. Next we choose an action. We initially explore selecting any of the three actions with equal probability. Then we either explore with probability ε by selecting one of the three actions, or exploit with probability $1-\varepsilon$ by considering the best action, i.e., the one that maximizes the reward in that state.

Once we know a state and action taken in that state, we can compute the reward obtained as described earlier. We can also obtain the next state of which only the ESD level is updated due to the action, while all the other dimensions of the 4-tuple can be obtained by considering the load, solar PV supply, day ahead commitment, real time cost, and real time credits. Using the current state, action, reward, and next state, we could update the Q-table using the following python syntax

```
Q[state][action] += alpha * (reward +
              gamma * np.max(Q[next_state]) - Q[state][action])
```

where `alpha` is known as the learning rate, `gamma` is known as the discount factor, and `np` is used to denote the python library called `numpy` which is already imported in the first few cells.

The term `Q[state][action]` is the updated entry of the Q-table, in particular corresponding to the entry in row "state" and column "action". The way it is updated is to use a learning rate `alpha` which is set to 0.1 in our examples, and it denotes the extent to which newly acquired information is used in updating the estimates. So, the weight given to new information is `alpha` while that of previous estimate is 1-`alpha`. The new information is the new reward plus a discounted benefit of future rewards. The discount factor is `gamma` and it pre-multiplies the best of all the actions once we are in the next state. We use a discount factor of 0.9, however this and the learning rate are tunable hyper-parameters.

In the training phase we go through multiple passes through the training data by sequentially updating. In our results we used 10 iterations or episodes, although one could experiment with other values. Once this is complete, we use the resulting Q-table to take actions in the test data. In each interval in the test data, we first obtain the state, and then look up the Q-table to see which action results in the maximum reward, and take that action. If we were to implement this, one of the issues is that over time in the test set, the optimal action results in the ESD getting drained completely. To avoid this, we use a penalty cost for discharging the ESD. Thus the reward in each state would be the usual reward minus a penalty for discharging from the ESD (equivalently reward for charging it).

We present two numerical examples. First is a continuation of the previous simulated example where we did a day-ahead analysis and obtained the day-ahead commitment. Here we will see what the results of the Q-learning based decisions reveal. To obtain the Q-table, here too we use the same 35-day training data as described earlier. To get day-ahead commitment for those, we had to first obtain in-sample forecasts for the training data and use that to go through a similar analysis of what was done in the testing data. This way we are able to iterate and obtain the Q-table for as many states as observed and the three actions.

(a) Actions at beginning of interval (2 - charge, 1 - do nothing, 0 - discharge)

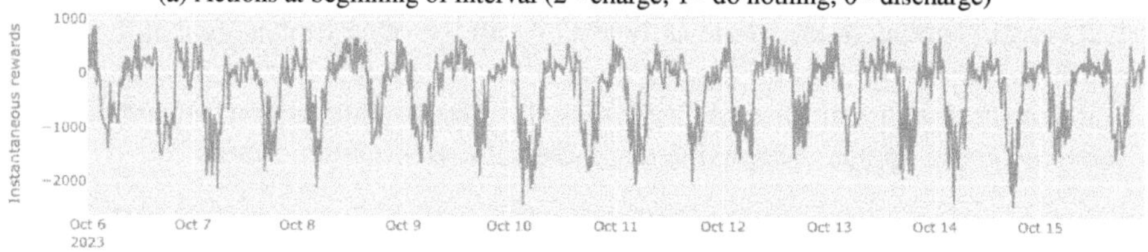

(b) Instantaneous rewards in each interval following actions taken

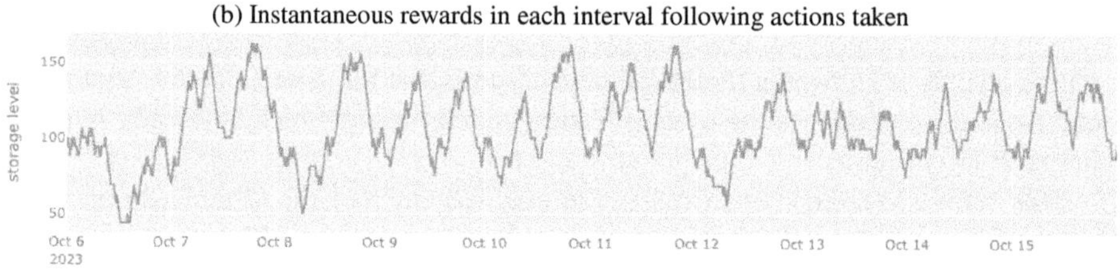

(c) Resulting storage level due to actions taken for all t and k in a day

Figure 5.5 Outputs for real time actions across 10 test dates: simulated data

Once the Q-table is trained using Q-learning from the training data, we consider the test data. Here in each interval we first obtain the state, then look up the action that maximizes the reward. The actions in each of the intervals is shown in Figure 5.5 (a) for all the 10 days. Notice that all three actions charge to the maximum extent (denoted by 2), do nothing (denoted by 1), and discharge to the maximum extent (denoted by 0) are all performed and also fairly regularly. The instantaneous rewards in each interval is in Figure 5.5 (b). It is important to note that the rewards are due to the amount purchased or sold to the utility grid, $X_{(t,k)}$, which depends on the demand, solar PV supply, day-ahead commitment, and the action $Z_{(t,k)}$.

Hence, it is entirely possible that even though our action is to discharge, our rewards may be negative. That is because the discharged amount is going toward the load while the amount of power committed is not sufficient. However, notice that there are two cycles of instantaneous rewards in each day and that corresponds to the net load cycle, the day-ahead commitment cycle, and the real time costs/credits cycle. Finally, refer to Figure 5.5 (c) which gives the ESD levels $U_{(t,k)}$. If we did not use a penalty for discharging from the ESD, this graph would quickly hit zero in the first two days and stay mostly at zero. Notice from the figure that when we continuously charge, the level goes up and vice versa. Also, the level is maintained above half the storage capacity.

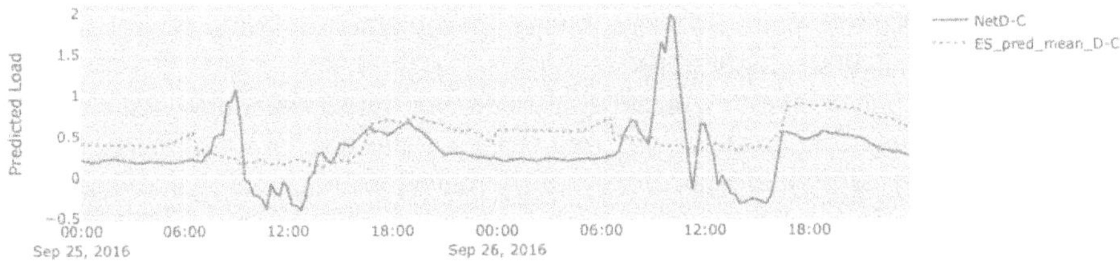

Figure 5.6 Net load: ground truth versus forecasted for real-inspired data

Real-Data Inspired Example

This is the second of the numerical examples. Clearly, it would be ideal to collect demand load data, solar supply data, day-ahead price data, real-time cost data, and real-time credit data from a single microgrid. However, since such a system does not exist, we take the next best approach which is to collect similar data from various sources. The goal was to maintain the level of uncertainty and variability in the data. We obtained 26 days of data starting September 1, 2016. The demand data was from households (`http://www.doc.ic.ac.uk/~dk3810/data/`) and solar PV supply data from NREL (`http://www.nrel.gov/midc/`). For the day-ahead prices, we used hourly clearing price data from CAISO for the same time period.

For the day-ahead costs, we averaged the 4 prices from the four markets (SDGE, VEA, PGAE, and SCE) to obtain an average day-ahead price. Then, we used 5-minute electricity prices from New England ISO (`http://iso-ne.org/`) to obtain 15-minute real-time costs and multiplied it by a random fraction to obtain 15-minute real-time credits. The household demand load and the solar PV power data were scaled suitably so that there is a significant amount of the load that is met by PV (including at times the solar power supply exceeding the demand). Also, the day-ahead prices had to be scaled so they were on average between the real time costs and credits.

Although such a mix and match is not ideal, it is still a worthwhile approach to understand how actions can be taken with the variability and uncertainty. Hence we call this as a real-data inspired example. Here the Pearson correlation coefficient was about 1.1-1.3% indicating that the net load (demand minus solar PV supply) is nearly independent of both real-time cost and real-time credit. See Figure 5.6 for the last two days of the test data where we compare the net load (blue) against the forecasted net load (red) at 4pm on the previous day for the whole day. Notice that during some daylight hours the net load is negative indicating a vast amount of solar PV supply.

Getting forecasts accurately 8–32 hours in advance for 15-minute granularity is extremely difficult using an exponential smoothing model. However, for that, Figure 5.6 does indicate a reasonable quality of forecast. The forecasts do take a dip in the daylight hours and a leap during dark peak hours. Notice that during some times the forecasts over-predict and at other times it under-predicts. However, the forecasted values are not nearly as variable as the ground truth. This is an important aspect. In some sense, the forecasted values are the means whereas the ground truth is a realization of a sample path. So there is very little chance it would be possible to make extremely accurate predictions.

However, we can do the day-ahead analysis using numerical values for some of the ESD parameters. In particular, we used charging and discharging efficiencies as $\eta_{dis} = 0.85$ and $\eta_{char} = 0.85$ respectively. Also, the values of charging and discharging capacities were equal to 0.025 (for c_{dis} and c_{char}). We set $\mathscr{K}$ to be 0.2 while we assumed u_{init} to be half that at 0.1. Using this we first compute the day-ahead commitment without storage and add to it the storage related adjustment to eventually obtain the day-ahead commitment. Using that and the Q-learning approach described above, we can obtain the optimal action to be taken at the beginning of each interval in the test set.

We train the Q-table using the training data which is the first 20 days and leave the last 6 days for testing. The results are described in Figure 5.7 for the last 6 days of testing. Figure 5.7 (a) shows the actions with 2 denoting charging, 1 denoting neither charge nor discharge, and 0 denoting discharging. Here there are longer periods of the same action than in the simulated data case. However, it is also important to realize that the 6-day dataset has points more spread out than the 10-day dataset. In Figure 5.7 (b) the instantaneous rewards are depicted. Notice that they are mostly costs with very little opportunity for big rewards although solar penetration is much higher.

However, the reason for low rewards might be that we are generally under-committing on day-ahead market as the real time costs are erratic. Also, the real-time credits are extremely low that any amount of sale only results in a moderate revenue. The sharp dip in instantaneous revenue on Sep 26 occurs at 10:15am which also corresponds to the sharp peak in Figure 5.6 on that day. So, to compensate for that surge in demand, a large amount of power is procured from the utility grid in real-time market. This causes a big drop in instantaneous reward. However, the play between the prices, day ahead plan versus real-time course correction, and inaccuracy of forecasts can drive varying behavior in a just small set of days.

Next let us focus our attention to Figure 5.7 (c). Here the effect of the actions on the ensuing level of energy stored in the ESD is presented. Around the middle of Sep 21, the ESD level goes to zero. Then around Sep 24 and again on Sep 26 the ESD level hits the maximum limit of $\mathscr{K}$. The Sep 21 corresponds to a large amount of discharging (from Figure 5.7 (a)). However on Sep 24 and Sep 26 the level lingers around $\mathscr{K}$ as the action switches between charge and discharge. Notice that the ESD level gradually increases and decreases although the other metrics show abrupt changes. Here too we avoid crashing to zero and staying there by explicitly using a penalty cost. This way the ESD is used as a reservoir for the time of need as opposed to a constant source of power.

Remark

Although we separately performed the day-ahead analysis and the real-time course correction, in practice this would be done together. A combined approach would

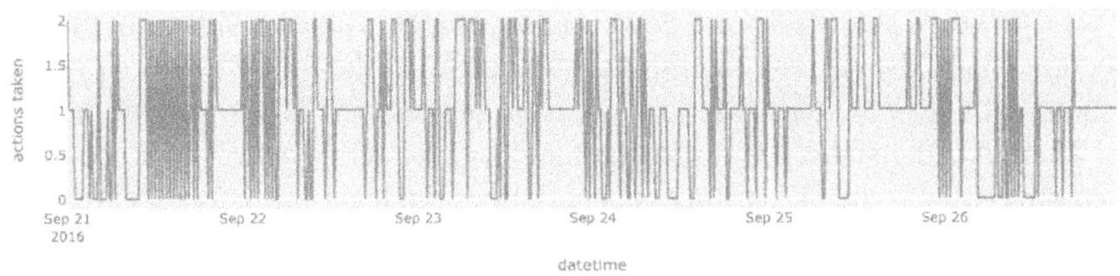

(a) Actions at beginning of interval (2 - charge, 1 - do nothing, 0 - discharge)

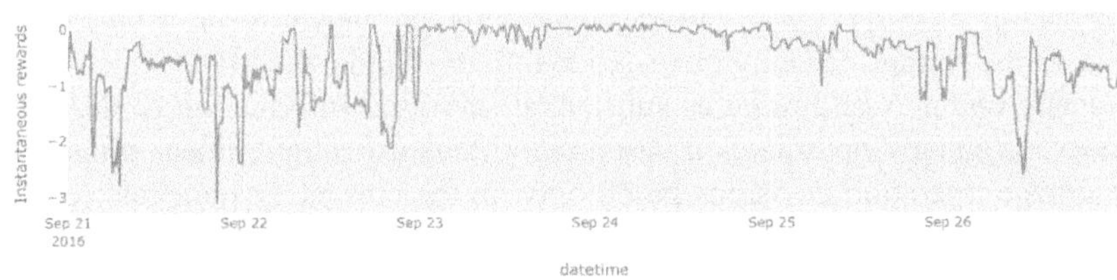

(b) Instantaneous rewards in each interval following actions taken

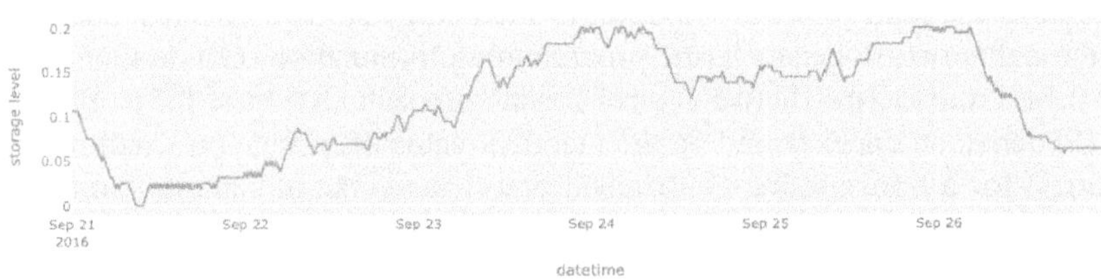

(c) Resulting storage level due to actions taken for all t and k in a day

Figure 5.7 Outputs for real time actions across 10 test dates: real-inspired data

require a similar training for both models and do a forecast in case of day-ahead. However, while implementing the results in an actual setting here is the process to be followed. On a given day at 4pm, all the day ahead commitments are made with a forecast of what u_{init} would be at midnight based on the current value at 4pm. This part has not been done here but is rather straightforward to do. Then we do the real-time predictions until 4pm the next day and then observe the storage level. At this time we not only predict u_{init}, but if needed adjust what may be expected at the end of the day such as more or less than u_{init}. One could also consider adjusting the penalty. In this way an iterative approach can be used and the model can be retrained from time to time.

5.4 A CASE FOR DISTRIBUTIONAL FORECAST

Before making a case for distributional forecasts, it is important to acknowledge other ways to solve such real-time power management problems shown in the previous section where the current conditions are known but the future evolves randomly. One such method is to use a Markov decision process (MDP) to model the future states. There too the states are discretized and a finite set of actions are considered. The key difference is the Q-learning approach does not learn the state transition probability matrix which the MDP is based on. Details of how to use MDP to determine actions in such real-time situations can be found in Kwon et al [29] and the references therein.

Another approach is to use Lyapunov optimization based on-line algorithms for the operation of ESDs (see Koutsopoulos et al [26], Huang et al [19], and Lakshminarayana et al [30]). These on-line algorithms are shown to be asymptotically optimal, as the storage capacity increases to infinity. Unlike the MDP, which is computationally complex and requires substantial statistical information of the system dynamics, Lyapunov optimization based algorithms use simple linear programs to make storage operation decisions based only on the current system state (e.g., the current storage level). For example, if one were to use the algorithm in Huang et al [19], it performs well when the storage capacity is significantly larger than the hourly maximum charging/discharging rates.

The joint scheduling of variable wind generation and energy storage systems is studied in a similar fashion that we saw for solar and storage. There the objective is to maximize the joint profit of wind farms and energy storage systems, and that is done through a two-stage stochastic programming formulation (Garcia-Gonzalez et al [14]), and a model-predictive-control-based approach (Xie et al [53]). Bose and Bitar [8] derive an upper bound on the marginal value of storage (at small installed capacities) for a transmission-constrained power network. In summary, managing systems with renewable sources (wind or solar) with storage has been well studied in the academic literature.

Another aspect that we have not looked at in this book is demand response where the load can be adjusted or shifted to a different time. Kwon et al [27] consider such a problem to mitigate intermittent renewable energy supply and volatile electricity price through demand response by solving two-stage stochastic mixed-integer program to obtain the optimal day ahead procurement. Data centers are good candidates for *demand response* since they have manageable and flexible workloads (see Wierman et al [52]). Motivated by that, Kwon et al [28] consider managing servers by powering them on and off as well as running them at different speeds as a way to perform demand response.

In all the above cases there is a need to be able to characterize the load, solar supply, and power as a stochastic process. Hence it only makes sense to make distributional forecasts as opposed to a point forecast. Further, in some of the calculations it is necessary to obtain distributional forecasts. For example, to determine the day-ahead commitments in Equation (5.2), we needed the CDF of the net load, not just the point forecast of the net load. That is a clear use case of distributional forecast.

Details on obtaining distributional forecasts is only in the next chapter. Here we simply show its importance in context.

In the previous chapter the day ahead forecasts had fairly large errors, there is only so much they can be improved. The main reason is that the errors are due to factors that cannot be measured at the time of making a forecast. However, if the use is in making decisions based on a distribution, perhaps the effort is in characterizing the distribution as effectively as possible. That said, next we present a somewhat similar example pertaining to day-ahead planning for a solar PV farm that makes a commitment and sells power to a utility grid. The objective is to make a case for distributional forecasts and go into detail of obtaining those forecasts in the next chapter.

SOLAR FARM SETTING AND APPROACH

In this section we consider a special type of microgrid which we alluded to earlier in this chapter. It is a solar farm that is large scale. It does not have internal demand and it does not have an ESD. All it does is goes into a contract to sell power to a utility grid. However, the utility grid goes into a pricing agreement with the solar farm, and the granularity is 15-minutes. Using the notation we had earlier let $Q_{(t,k)}$ be the day ahead power commitment (in MW) promised by the solar farm to the utility for all $(t,k) \in \mathscr{A}_{test}$. Clearly, when there is little or no sunlight $Q_{(t,k)} = 0$ as there is not reason to commit to any power. It is worth reiterating that this commitment is made a day in advance.

There is a flat revenue of R dollars per MWh for the solar farm from the utility grid. So in the 15-minute interval (t,k), if the solar farm commits to providing $Q_{(t,k)}$ MW of power a day ahead, then it receives a revenue of $RQ_{(t,k)}/4$ dollars (the division by 4 is because it is $1/4$th of an hour). During the 15-minute interval in real-time, suppose the realized solar generation (i.e., the amount the solar farm can supply) is a random variable $S_{(t,k)}$ MW. If the realized generation is greater than $Q_{(t,k)}$, then the solar farm curtails the excess power (or gives it to the utility for free). Either case, from the solar farm's standpoint there is no revenue from the extra power generated.

The tricky part in the contract is that if the realized generation $S_{(t,k)}$ is less than $Q_{(t,k)}$. Now the solar farm falls short of its promise but the utility has already made plans a day ahead for the power. So the contract stipulates that the solar farm is charged a penalty of C dollars per MWh short of the committed amount. So, the revenue in that 15-minute interval would be $RQ_{(t,k)}/4 - C(Q_{(t,k)} - S_{(t,k)})/4$. This penalty C is for the utility grid to buy the remaining power from a more expensive source in the last minute at a cost C dollars per MWh.

From a logical standpoint $C > R$ otherwise: (i) the last minute provider would not be more expensive which would mean the utility grid would not have considered the solar farm as an economic source of power; (ii) the solar farm would be incentivized to promise as must as possible to maximize its revenue. In this arrangement, the utility grid spends $RQ_{(t,k)}/4$ and receives power of $Q_{(t,k)}$ MW during the 15-minute interval (t,k), whether it is from the solar farm or a combination of solar farm and an expensive last-minute supplier. For the utility, there is no decision to be made. Also, notice that R and C are fixed for all intervals.

The solar farm has to make a decision about what day-ahead power $Q_{(t,k)}$ MW it should commit to providing the utility grid. One option is for the solar farm to commit $Q_{(t,k)} = \mathbb{E}[S_{(t,k)}]$ which is the best point forecast that can be made a day ahead of the amount of solar power that can be generated. Are there other options? A reasonable objective for the solar farm is to maximize its expected revenue. To compute that, suppose we know the distribution of $S_{(t,k)}$ (we will describe how to get this in the next chapter). Let $F_S(x) = \mathbb{P}\{S_{(t,k)} \leq x\}$ be the CDF.

We can solve this problem independently for each (t,k) and hence we drop the (t,k) from the analysis with the understanding we are trying to maximize the expected revenue for a specific (t,k). Suppose we wish to commit a quantity Q, then the expected revenue (see reasoning provided above) in that interval is:

$$J = \mathbb{E}\left[RQ/4 - C\min(Q-S,0)/4\right]$$

with the understanding that J, Q, and S would have a subscript (t,k) which has been dropped, as stated above. We explicitly write down J in terms of the CDF of S, $F_S(\cdot)$ as

$$J = \frac{RQ}{4} - C\int_0^\infty \frac{\min(Q-x,0)}{4} dF_S(x)$$

by writing down the random variable conditioning on its realization as x.

We can rewrite the above equation for J as

$$J = \frac{RQ}{4} - \frac{C}{4}\int_0^Q (Q-x)dF_S(x)$$

which can be rewritten and grouped as

$$J = \frac{RQ}{4} - \frac{C}{4}\int_0^Q QdF_S(x) + \frac{C}{4}\int_0^Q xdF_S(x)$$

with the first integral relatively easy to compute. Hence we have

$$J = \frac{RQ}{4} - \frac{CQ}{4}F_S(Q) + \frac{C}{4}\int_0^Q xdF_S(x)$$

with the understanding that $F_S(0) = 0$. Since we wish to find Q that would maximize the revenue, we simply take the derivative of J with respect to Q. That results in

$$\frac{dJ}{dQ} = R/4 - CF_S(Q)/4 - CQf_S(Q)/4 + CQf_S(Q)/4$$

where $f_S(x)$ is the probability density function (PDF), i.e., the derivative of $F_S(x)$ with respect to x.

Notice that the second derivative

$$\frac{d^2J}{dQ^2} = -Cf_S(Q)/4$$

is negative for any Q as the PDF is always non-negative. Hence the optimal Q where

$$\frac{dJ}{dQ} = 0$$

corresponds to the maximum. Rearranging the terms we have the optimum power to commit to the utility grid Q^* satisfies

$$F_S(Q^*) = \frac{R}{C}$$

and since $R < C$, the right hand side can correspond to a probability. Hence

$$Q^* = F_S^{-1}(R/C) \tag{5.10}$$

which can be computed if the CDF is known. The details will be provided in the next chapter but we use a simple approach here for a numerical example which we consider next.

NUMERICAL EVALUATION

Here we consider a data set from Elia [6] which continuously tracks and forecasts solar power generation in Belgium. Elia publishes the results on their website and it is free to download. The data is in 15-minute granularity. We use data for all the dates in October 2019 to December 2019. The dataset is across the entire country of Belgium and hence is not a solar farm. However, what makes the data set ideal for this study is that it has day-ahead point forecasts as well as the ground truth of the actual solar power produced. Although we only use the data and time column, the day-ahead forecast, and the actual solar power generated, the data set has many other columns such week-ahead forecast and day-ahead forecast at 11am, among others.

For times when there is no sunlight or the power is extremely low, we simply commit to 0. We only use the remaining data after filtering out the zero values. The MAPE for the day-ahead forecasts is 32.6% which is expected as day-ahead forecasts are not as accurate. We do not make any forecasts and simply use the data provided. In particular, we assume that the day-ahead forecasts correspond to the estimated mean of the ground truth solar power generated in actuality. However, to obtain a distributional forecast we need more information. For that we use data from October and November to train and test using the December data. This is a rather simplistic approach, better methods would be discussed in the next chapter. Here we simply estimate and do not do any time-series forecast.

We consider $R = \$40$ per MWh and $C = \$46$ per MWh as the revenue and penalty cost respectively. This information is not given in the Belgian site. We consider three policies: *Policy 1* - we simply use the day-ahead forecast as the day-ahead commitment; *Policy 2* - we use Equation (5.10) to determine the commitment where F is the CDF of a normal random variable with mean given by the day-ahead forecast, and variance that we will predict; *Policy 3* - we empirically obtain the percentile (equal

Table 5.2
Total revenue in test data due to the 3 policies

	Policy 1	Policy 2	Policy 3
Net Revenue	\$2,803,303.57	\$4,871,304.13	\$5,189,977.30

to that in Equation (5.10)) using the training data by carefully scaling against the day-ahead forecast. We explain the methods next, followed by the results.

Suppose that we are considering a specific hour and 15-minute interval in the test data, say 11:15am. Then we consider all the 11:15am data in the training date (in this data set there would be $n = 61$ data points in total in the October and November training data). We compute the mean square error between the actual solar power and the day-ahead forecast (which we assume is the mean) and use that as an estimate of the variance. We use the mean and variance in the Normal distribution in Policy 2. For policy 3, we use the same $n = 61$ training data point for 11:15am but a different approach. First notice that R/C is roughly 87% and we want to use the 87th percentile. However, the use is a bit more nuanced.

We first use the $n = 61$ data points, sort them, and obtain the 87th percentile value. Then we scale the 87th percentile value using a multiplier that is the ratio of the day-ahead forecast of the test data to the mean of the ground truth of the training data. Understandably, there are many other ways of doing a scaling and getting a percentile. In fact, we recommend that the approaches in the next chapter be used. Once we have the day-ahead commitment for each of the three policies, we use the 31 test days and each of the 96 intervals (of which most would be zero due to no sunlight in December) (t,k) to compute $RQ_{(t,k)}/4 - C\min(Q_{(t,k)} - S_{(t,k)}, 0)/4$ and sum over all (t,k) to get the total cost as shown in Table 5.2.

The important take-away from Table 5.2 is that Policy 1 which uses just the day-ahead forecast as the day-ahead commitment results in a much lesser net revenue. However, it is astounding how often in practice one simply uses the best estimate of the forecast directly in the model. This itself is hopefully a good indicator of the need for distributional forecast. Although in this case Policy 2 fared worse than Policy 3, that does not always happen. This data set was far from a normal distribution. In fact even the gamma distribution was tried but that did not yield as good a result. In the next chapter we will consider parametric (like policy 2 from a specific distribution) and non-parametric methods (like policy 3 where we obtain percentiles without considering any specific distribution).

6 Distributional Forecasts

In many systems, especially in forecasting solar power in day ahead scenarios, there is a large chance of a significant error. This error can also be for other systems with significant amount of unknown reasons for their values to fluctuate. Thus we are talking about not just solar supply, but also load demand processes and costs. In fact, even sources with changing amount of solar panels, due to newer panels additionally being brought in, also have this. In fact, this theory is amenable to any system which is difficult to predict over time. However, we stick to the solar forecasts case, and allow us to consider the new panels that are an added case.

In that light, on one hand this problem is a simple extension of what we have seen thus far where we consider forecast intervals as opposed to point forecasts. Thus instead of a single point, we will be analyzing a range of values. On the other hand the key difference is that none of the values obtained are observed. In other words, there is no realization of the results or effect of training that goes on. Thus unlike the previous case where we say we have the ground truth, here we do not have it. That is a significant difference, although there is reason to believe that is the case as one can never be sure of a random act.

Further, the results are not independent each time. In other words, the reason the observed value could be lesser than average could be the case for multiple periods. Thus the errors are not necessarily random every time. This is true of actual processes too, for example, if there is a huge cloud, the amount of solar is going to be lower than what is observed for a large number of observations. For that reason, it is not enough to take samples over a small range, but they have to be taken for a large number of values. Thus, it becomes extremely important to collect a lot of samples over a large number of values.

That said, it is extremely important to realize that although this is the last chapter of this book, it is one of the most realistic ones. The predictions are real and conform to true happenings. It is particularly useful when the random quantities could take very different values, depending on the circumstances. While there are two interpretations of that which is that many of the unknowns are not mapped is one way. But there is another way which says we can assume that all that we can measure are mapped. It is only the random realizations that will be different. Hence the study is from a different aspect which we will see here.

6.1 INTRODUCTORY REMARKS

In probabilistic forecasts, we say something like *there is a 60% chance of rain tomorrow*. Notice that we are essentially forecasting the probability it will rain tomorrow. So the event is about rain but we do not know whether it will rain. All the sentence

DOI: 10.1201/9781003403500-6

says is that there is a 60% chance that it will. So that means if we were to compare across multiple days of 60% chance of rain, on 60% of the days it would rain and on 40% of the days it would not. While this is the most fundamental of discrete quantities to be forecasted and is complicated, we extend it a step further and consider continuous quantities.

For example, the amount of rain could be the uncertain quantity to be estimated. In other words, a distributional forecast is essentially a probabilistic forecast for continuous variables such as the amount of rain. For instance, we may be interested in the probability of getting more than 6 inches of rain tomorrow. For example, if the amount of rain is 6.1 inches, you might say it is close, but a value of 12.3 inches may be considered as extremely far. But it may just be a chance that 6.1 inches might be more accurate than 12.3 inches in this one case, So it becomes important to be predicting a number of values with different probabilities.

Thus, what we are interested is in obtaining the distribution of the number or inches of rain we would see tomorrow. If the chances of rain is nearly zero, then with a very high probability the distribution of rain would be zero. But when the chances of rain is nearly 100%, we could say the probability that it would rain for more than 3 inches is 0.5, more than 4 inches is 0.3, and so on. However, what about getting more than 3.5 inches? Is that more than 0.4? So, although it would be ideal to get the entire CDF of the random variable, it is not practical. Hence we strike a balance between the number of intervals we would predict.

Before describing that, it is worthwhile noting that we would, for the purpose of this text, be interested in predicting the amount of solar power during a period of 15 minutes, say. If the solar irradiance is 800 kW/m^2 right now. Then in the next 15 minutes, depending on the type of day it is, we might say that it would be less than 850 kW/m^2 with probability 0.8, it would be less than 650 kW/m^2 with probability 0.5, and it would be less than 250 kW/m^2 with probability 0.3, for example. Of course, a lot of this depends on the cloud conditions. On a perfectly sunny day, the solar irradiance could be between 820–825 kW/m^2.

Likewise on an overcast day, it might be 200–250 kW/m^2 with a high probability. But on a partly cloudy or cloudy day, it might have a larger width, all the way from 250 kW/m^2 to about 850 kW/m^2. We wish to be able to predict this as well as characterize this. While mathematically, it is necessary and accurate, the downside is there is no ground truth available. This makes is difficult to have one model that is extremely useful. But, we can use the methodology here to say which of the two or more models would fare better under the given circumstances. This is quite effective in the study of such distributed forecasts.

We now represent this using random variables and distributions. Let Y be a continuous random variable with CDF $F(y) = \mathbb{P}\{Y \leq y\}$ for all $y \in \mathbb{R}$. Clearly, as y increases, so does $F(y)$. However, the values Y at a given time is different for different days of the week and also depends heavily on the type of day it is. Also, we don't typically know $F(y)$. So one of the jobs is to see if $F(y)$ observed is the same as that forecasted. However we only observe one value of y and so the question becomes what value of y and $F(y)$ were observed and predicted, knowing both are unknown. Thus we take a small set of $F(y)$ to make sure they match.

For that, although $F(y)$ takes all real values in $[0,1]$ for various y, many times we are interested in specific values called quantiles. Popular quantiles are quartiles (0.25, 0.5, and 0.75), deciles (0.1, 0.2, ..., 0.9), and percentiles (0.01, 0.02, ..., 0.99). We in this book are typically interested in the middle one as the first one is too broad while the last one has too many intervals. Without knowing what value $F(y)$ takes for the observed y, we compute the $F(y)$ that best represents the choice considered. Of course, it is much more effective in comparing two different $F_1(y)$ and $F_2(y)$ to determine which is better.

In the next few paragraphs we explain various aspects with respect to $F(y)$ subsequently that is predicted. We first consider representing quantiles. In some situations (especially in news vendor-like buffer-less and day-ahead problems) we may be interested in a specific quantile. We call that the *target quantile* τ such that $F(\cdot) = \tau$. For example, when $\tau = 0.5$, $F^{-1}(\tau)$ is the **median**. In the day-ahead commitment,

$$F(q^*) = \frac{r-d}{r-c} \quad \text{or} \quad q^* = F^{-1}\left(\frac{r-d}{r-c}\right)$$

essentially $\tau = \frac{r-d}{r-c}$ and $F^{-1}(\tau)$ is the quantity of power to commit as described in Section 5.2.

In other situations we may be interested in the entire spectrum (like decile) and not just a single value such as τ. This is especially the case we are looking forward to here in this section. In particular, we wish to know if we have an $F(y)$ for various values of y. In many industries it is common to use the 20th (represented as P_{20}) percentile or 80th (represented as P_{80}) percentile for planning, In other words, we wish to computer essentially $F^{-1}(0.2)$ which is P_{20} or $F^{-1}(0.8)$ is P_{80}. Thus from a forecasting standpoint we may be interested in predicting the entire decile, i.e., P_{10}, $P_{20}, P_{30}, \ldots, P_{90}$, at all times.

This is known as the decile predictions at the quantile that we are interested. This way we are interested in a large number of percentiles but not all 99 from 1 to 99. Actually one could even do more than that, as a matter of fact. The reason for keeping it small (but not too small) is that it is manageable on one hand which on the other hand it does not go out of control. By out of control we mean the number of samples we need to take to test all these. If we have all 99 values $P_{01}, P_{02}, \ldots, P_{99}$, then so many terms need to be tested. Recall that for each 15-minute interval you have a different function $F(y)$. Thus computing it for 9 values seems a bit more reasonable.

Next we explain a few more comments on quantiles and then explain what we would do. Consider two random variables Y and Z so that Z is an affine transformation of Y

$$Z = aY + b$$

for some constants a and b. It is well known that $\mathbb{E}[Z] = a\mathbb{E}[Y] + b$, thus preserving the affine transformation. Likewise, quantile inverses also preserve affine transformation, for example, the P_{40} of Z is essentially a times the P_{40} of Y plus b as

$$\mathbb{P}\{Y \le P_{40}\} = 0.4 \quad \Longrightarrow \quad \mathbb{P}\{Z \le aP_{40} + b\} = 0.4$$

where $P_{40} = F^{-1}(0.4)$. Note that even standard deviation does not have this property, unless $b = 0$.

Thus far we have only mentioned the properties of the quantiles. We still have to obtain $F(y)$. That way to predict $P_{10}, P_{20}, \ldots, P_{90}$, or for that matter any $F^{-1}(\tau)$ we need methods to compute $F(y)$. For that in Section 6.3 we will present a parametric approach and in Section 6.4, a non-parametric method. However, to obtain the $F(y)$ in case of non-parametric approach and for both approaches to obtain ways to compare various methods, we first consider the pinball loss function which can be used to compare multiple algorithms in Section 6.2 and determine the winner.

6.2 PINBALL LOSS FUNCTION

Say that we determine distributional forecasts of hourly demand, supply, and real-time price of grid power. For example, an algorithm predicts the quantile levels $(F^{-1}(0.1), F^{-1}(0.2), \ldots, F^{-1}(0.9))$ for each of the values: hourly demand, supply, and real-time price. The main question then becomes how good are these forecasts. The major complication is that we do not have any ground truth for this. But we need to determine if that choice is better or another choice, say, $(G^{-1}(0.1), G^{-1}(0.2), \ldots, G^{-1}(0.9))$ is a better forecast. The most straightforward idea is to say that the one giving us the smaller error is the better one.

The way we go about doing that is using a pinball loss function. For the sake of simplicity, we only consider the case of solar power supply. Although it is pretty straightforward to extend it to any system whose distributional forecast is needed. Recall that there is no ground truth to test the distributional forecasts against (since we only get one realized value and not a collection). However, we do have a single sample value for all cases. But, we do not know if the single sample value corresponds to what P_x value. Hence, we consider it to be true for some value such that a fraction x times it is expected to be smaller than P_x.

Recall that we do not know $F(\cdot)$. Hence we need to be able to test multiple algorithms using the said function. For that, the pinball loss function is an excellent choice. We define the pinball loss function as $L_\tau(y,z)$, where τ is target quantile, y is observed value and z is quantile forecast such that

$$L_\tau(y,z) = \begin{cases} (y-z)\tau & \text{if } y > z, \\ (z-y)(1-\tau) & \text{if } y \leq z. \end{cases}$$

Essentially, z is a prediction of $F^{-1}(\tau)$ so that a fraction τ time $y \leq z$.

Of course, since $L_\tau(y,z)$ is unknown when $F(\cdot)$ is not known, we try to compute $F(\cdot)$ that results in the smallest pinball loss function. However, when $F(\cdot)$ is known and it is used to determine the value of y, then the best choice of z is indeed $F^{-1}(\tau)$ as that minimizes $L_\tau(y,z)$. That is the first result that we show and in the process explain for unknown $F(y)$ cases, the best option is indeed $F^{-1}(\tau)$. We next explain the rational behind that and use a proof technique to describe it. For that we assume that Y is a random variable with CDF $F(y)$ for all y.

In particular, let Y be a continuous random variable with CDF $F(y) = \mathbb{P}\{Y \leq y\}$ for all $y \in \mathbb{R}$. Let τ be an arbitrary target quantile for which we wish to predict z

before the sample y is drawn from Y. Recall the pinball loss function ($L_\tau(y,z)$ where τ is target quantile, y is observed value, and z is quantile forecast)

$$L_\tau(y,z) = \begin{cases} (y-z)\tau & \text{if } y > z, \\ (z-y)(1-\tau) & \text{if } y \leq z. \end{cases}$$

Note that in the above situation although we know y, we need to predict z before the actual value of y is known.

The main implication is that if we are an Oracle that knows the CDF $F(\cdot)$, then the optimal choice of z would be to predict it as $F^{-1}(\tau)$. But in practice we do not know the CDF, so we select the algorithm that minimizes the sample average of all the test predictions across all τ. With that understanding, as Y is a continuous random variable, we have,

$$\mathbb{E}[L_\tau(Y,z)] = \int_{y=-\infty}^{\infty} L_\tau(y,z)dF(y)$$

as z is predicted before y is known. The reason is that z is forecasted value in advance while y is the actual value. As stated earlier, Y has the ability to lay down any value in its range.

Since $L_\tau(y,z)$ takes different values depending on whether y is greater than or lesser than z, we can write down

$$\mathbb{E}[L_\tau(Y,z)] = \int_{y=-\infty}^{z} L_\tau(y,z)dF(y) + \int_{y=z}^{\infty} L_\tau(y,z)dF(y)$$

by considering two cases, one where y is bigger than z and other when y is smaller than z. By writing down the corresponding expressions, we get

$$\mathbb{E}[L_\tau(Y,z)] = \int_{y=-\infty}^{z} (z-y)(1-\tau)dF(y) + \int_{y=z}^{\infty} (y-z)\tau dF(y)$$

by simply writing down the cases of $L_\tau(y,z)$ when $y > z$ and when $y \leq z$.

Combining the like terms we get

$$\begin{aligned} \mathbb{E}[L_\tau(Y,z)] &= (1-\tau)z\int_{y=-\infty}^{z} dF(y) - (1-\tau)\int_{y=-\infty}^{z} ydF(y) \\ &\quad +\tau\int_{y=z}^{\infty} ydF(y) - \tau z\int_{y=z}^{\infty} dF(y) \end{aligned}$$

by writing down terms with z and those with y with the understanding that z is a constant fixed value. Now, we ask the question if $z = F^{-1}(\tau)$ minimizes the function $\mathbb{E}[L_\tau(Y,z)]$.

Let us denote (for convenience of taking derivatives) $\psi(z)$ as

$$\psi(z) = \mathbb{E}[L_\tau(Y,z)]$$

so that we have a simple expression to denote in our results. Notice that z is the only choice that is varying, and we need to check if choosing $z = F^{-1}(\tau)$ minimizes the function $\mathbb{E}[L_\tau(Y,z)]$. From the previous expressions and performing integration for the simpler expressions involving $\int dF(y)$, we get

$$\begin{aligned}\psi(z) &= (1-\tau)zF(z)-(1-\tau)\int_{y=-\infty}^{z} ydF(y) \\ &\quad +\tau\int_{y=z}^{\infty} ydF(y)-\tau z(1-F(z))\end{aligned}$$

where the first and last expression is obtained after computing the integral.

Taking the derivative of $\psi(z)$ with respect to z, we get

$$\begin{aligned}\psi'(z) &= (1-\tau)F(z)+(1-\tau)zf(z)-(1-\tau)zf(z) \\ &\quad -\tau zf(z)-\tau(1-F(z))+\tau zf(z) \\ &= F(z)-\tau\end{aligned}$$

where the probability density function of Y is $f(y)=\frac{dF(y)}{dy}$. It is worthwhile noticing that to get the above expression, we use the fact that the derivative of

$$\int_{y=-\infty}^{z} ydF(y)$$

with respect to z is $zf(z)$ which is a standard result in calculus.

We showed that

$$\psi'(z)=F(z)-\tau$$

and hence the derivative of $\psi(z)$ is a very simplistic relation. Then, setting $\psi'(z)=0$, we get

$$F(z^*)-\tau=0$$

where z^* is either the maxima or the minima. We can write down $z^*=F^{-1}(\tau)$ based on the expression $F(z^*)-\tau=0$. We still have to show that it is the minima. For that, since for all z,

$$\psi''(z)=f(z)\geq 0$$

where the left hand side is simply the derivative of $\psi'(z)$ with respect to z giving us the probability density function that we defined earlier, and the right side equality is just a function of the probability density function, which by definition much be greater than or equal to zero.

Hence, we know z^* is a minima point. Also, since $F(y)$ is a non-decreasing function, z^* is a unique minimum. Thus, the value of z that minimizes

$$\mathbb{E}[L_\tau(Y,z)]$$

is $F^{-1}(\tau)$. Of course, this assumes that $F(y)$ is known and the result was derived. Now, if $F(y)$ is unknown, but we have multiple values of y available. Then among multiple values y_1, y_2, ..., y_n, the value of $G(\cdot)$ that is the best is the one that minimizes $\sum_{i=1}^{n} L_\tau(y_i,z)/n$ where $z=G^{-1}(\tau)$. This is an approximation for $\mathbb{E}[L_\tau(Y,z)]$ by considering n samples (as opposed to $n=\infty$). This also assumes that for all n cases the $F(y)$ is a constant, though unknown. We will subsequently relax this but we first show how this is true.

Table 6.1
Training data of y values (rounded off to two decimals)

0.67	4.77	0.09	0.04	0.87	1.52	1.34	5.71	2.05	1.03
0.41	0.02	0.02	2.49	1.05	0.66	1.85	0.30	0.13	1.72
2.43	0.37	0.41	1.42	0.03	1.94	0.88	0.19	2.30	0.03

SAMPLES FROM AN IID DISTRIBUTION

As described in the previous paragraph, we consider a sample of $n = 30$ values of y which is the training data to obtain the value of z for which $\tau = 0.7$. This is shown in Table 6.1. The first approach is to simply take the P_{70} from this data which is 1.52, if we were to sort the data (that means 70% of the data are less than or equal to 1.52) which is the case indeed. Hence in this first method, we use $z = 1.52$ as the 70th percentile of the training data. The second approach computes the mean and standard deviation of the data as 1.22 and 1.36, respectively. Then, it computes the parameters of a Gamma distribution by obtaining the shape parameter as $(1.22/1.36)^2$ and scale parameter as $(1.36^2/1.22)$.

Now, it is worthwhile first talking about the choice of Gamma distribution. Notice that since the standard deviation is much higher than the mean, there are only a few distributions that would work. Although ideally one would have to perform a goodness of fit test to be sure the data fits. Since this is a relatively straightforward use case, where we are trying to illustrate the concept, that is not done. At any rate, the value of z corresponding to the P_{70} is 1.4452 (which equal the probability that the inverse of the CDF for a gamma distribution at 0.7 under the aforementioned values of shape and scale parameters). Notice the values of P_{70} are close in the first and second methods.

The third approach is not realistic but is considered as a benchmark for this data. Here we know the underlying distribution of the data that was used to generate this. In particular, this was generated from a Gamma distribution with shape parameter 0.5 and scale parameter 4. Thus we can immediately write down the P_{70} as 2.1484 which can be garnered without even having the training data. That makes sense since we know the method of generating the values randomly. For this data we have P_{70} of 2.1484 which is quite different from those of the first two methods. We shall see next how it impacts the test data.

In our test data in Table 6.2, we compute $\sum_{i=1}^{n} L_\tau(y_i, z)/n$ for the three methods using three different values of z. For that we first take for each y_i and compute $L_\tau(y_i, z)$ using $(y_i - z)\tau$ if $y_i > z$ or using $(z - y_i)(1 - \tau)$ if $y \leq z$ for each of the three values of z. This results in the values of y_i versus the approaches 1, 2, and 3. Then we take the average of each of the three columns to get $\sum_{i=1}^{n} L_\tau(y_i, z)/n$ as 0.9509, 0.9609, and 0.9025 for methods 1, 2, and 3. Here the method 3 is the smallest, while method 1 is the next, and method 2 is the largest.

Table 6.2
Testing data of y values (rounded off to two decimals) considered under three approaches

$L_{\tau}(y_i, z)$	$z = 1.52$	$z = 1.4452$	$z = 2.1484$
y_i	Approach 1	Approach 2	Approach 3
2.31	0.5499	0.6023	0.1101
1.08	0.1319	0.1094	0.3204
2.56	0.7260	0.7783	0.2861
0.04	0.4451	0.4227	0.6336
6.31	3.3501	3.4025	2.9102
0.01	0.4522	0.4297	0.6407
1.80	0.1949	0.2473	0.1050
1.13	0.1182	0.0958	0.3067
1.03	0.1472	0.1247	0.3357
1.02	0.1502	0.1277	0.3387
0.31	0.3632	0.3408	0.5517
5.88	3.0504	3.1027	2.6105
0.18	0.4033	0.3809	0.5918
0.47	0.3145	0.2921	0.5031
2.12	0.4179	0.4702	0.0094
6.04	3.1664	3.2188	2.7266
3.73	1.5455	1.5978	1.1056
1.16	0.1075	0.0851	0.2960
13.10	8.1037	8.1561	7.6638
2.47	0.6665	0.7188	0.2266
0.13	0.4160	0.3936	0.6045
0.30	0.3657	0.3432	0.5542
0.07	0.4339	0.4115	0.6224
1.86	0.2401	0.2924	0.0856
1.75	0.1630	0.2154	0.1186
0.28	0.3729	0.3504	0.5614
3.43	1.3343	1.3866	0.8944
0.22	0.3901	0.3677	0.5787
0.93	0.1772	0.1547	0.3657
0.75	0.2295	0.2071	0.4180
$\sum_{i=1}^{n} L_{\tau}(y_i, z)/n$	0.9509	0.9609	0.9025

It is important to mention that the values could change for a different set of 30 values taken in the data set and this is just a sample example. Of course, the unrealistic approach 3 is likely to perform the best but it is worth noticing that it is not too much different from approaches 1 and 2. Notice how close they are, of course that is expected since the two had relatively closer z values. The point here is to show that we need a way to obtain the z values using training. But in practice we are never going to actually have the same IID distribution for all the data points. This we consider in the next example.

SAMPLES FROM A SET WITH FIXED τ VALUES

The previous example was a good case to show how one could use the case of a single distribution for all values. However, a more realistic example is the case where each sample is from a different distribution. The objective is still to compute the pinball loss function, and the details of how one goes about selecting the CDF $F(\cdot)$ will be dealt with later. We will simply select a few options, and not the best option necessarily. Also, the approach used is far from the optimal solution. But the main point is to showcase the pinball loss function for various values. The important thing is to realize that it is for a specific dataset on a particular month.

Here we consider the dataset from Elia [6] which is exactly the same as that in the very end of Section 5.4. However, the setting here is different, and hence it would not be ideal to search there. However, the data set itself is the same (it is the use and the values of R and C that are different). For example, here too we consider 15-minute granularity data for the months October 2019 to December 2019. Using the day-ahead forecast available, we wish to make the 90th percentile of forecast. However, it is important to realize that the pinball loss function not only penalizes the case when $\tau \neq 0.9$ but also for how much away they are.

Recall the data is only for the time when there is some sunlight. In other words, times when there is no sunlight is completely discarded. For the remaining times we assume that the day-ahead forecast is a good starting point in this stage. So we do not make any forecasts but simply use the day-ahead forecast values. However, we still have to train the models to select a set of parameters that are useful in determining the variability in the values of the forecasts. So, we use the data from October and November for training, and use the data in December for testing. We consider 5 methods in total to obtain $F(\cdot)$, although we do not know which one would work.

For the first method, we simply use the day-ahead forecast itself at each hour as the 90th percentile of forecast. So in other words in the first method, the value of z is not even a function of τ. In other words z depends only on the value of the day-ahead forecast but not on τ. So $G_1^{-1}(\tau)$ is simply the day-ahead forecast. Now, for the second and third methods, we assume that the actual value is sampled from a mean equal to the day-ahead forecast and the standard deviation. The day-ahead forecast is available for the test data. But how do we get the standard deviation? For this we use the training data.

In the training data we compute the variance in the data between the actual values and the day-ahead forecasts for each 15-minute interval. Since we have 61 data points for each time of the day, we can compute the average of the squared time between the day-ahead forecasts and the actual values in the train data. Then we take the square root of the average and use it as the standard deviation estimate, assuming that the means occur at the day-ahead values (this is not true, but we will ignore it for now). Now, to get z values between the second and third methods there is a difference and we explain that next. Notice that neither method assumes homoscedasticity, but rather that the variance changes from time to time.

In method 2, we assume that the difference between the actual values and the day-ahead forecast is according to a normal distribution. While in method 3, we assume that this difference is according to a Gamma distribution. So in method 2, $G_2^{-1}(\tau)$ is the inverse of a normal distribution with mean equal to the day-ahead forecast in the test data and standard deviation equal to the value corresponding to the square-root of the variance at that time in the training data. In method 3, we use the inverse of a Gamma distribution with shape parameter equal to the variance divided by the mean while the scale parameter equals the mean divided by the shape parameter.

In method 4, for each 15-minute interval, we group all October–November historical values for that time. Then we compute the empirical percentile corresponding to τ. Then we scale this quantile by a normalization factor to match the day-ahead forecast (used as the mean in other methods). This anchors method 4 around the day-ahead forecast (like the other methods) but allows for non-parametric, multi-modal, and skewed behavior — exactly as needed given the multimodal plots. Group training data by time of day and store the empirical distribution of corrected measurements. For each December time stamp, we get the empirical distribution for that time. Then we compute its quantile. Normalize it so the mean of the empirical distribution aligns with the day-ahead forecast. We use this scaled quantile as the committed amount.

In method 5, the approach is similar to method 4 but the scaling is done first. For each 15-minute interval in the training data, we compute the ratio of the actual versus the day-ahead prediction for all training days. Then we take the empirical τ prediction of this ratio distribution. Then for each test date and time, we retrieve the percentile ratio for that time slot. Then we multiply by the test day's day-ahead forecast to get the value of z. This naturally anchors to each day's forecast and captures relative error patterns. So, if a whole day is cloudier or sunnier than average, absolute historical patterns may not fit but the above ratios might. Also, since days in December have naturally lower forecasts, the ratios would automatically adjust.

Note that in all the methods, we use a generic τ but it is relatively straightforward to use a specific value such as 0.9 which is what we will do. Then we compute the pinball loss function where y is the actual value and z is the value obtained using the five methods. Based on that, the pinball loss functions for $\tau = 0.9$ are described in Table 6.3. Notice that pinball loss function is certainly a function of the number of values above the actual, and this should be 0.9 ideally. The method 5 comes closest among all, and hence has the smallest pinball loss function.

Method 3 which boasted the second smallest pinball loss had reasonable values for the number of z values being larger than y as well as the variability in values z takes. In comparison, Method 4 although it has the exactly the same fraction of values greater than y, the values in the standard deviation is much larger in 4. Hence it has the worst value. In fact, method 1 although has a fairly small number of z larger than the actual y, what makes it have a small pinball loss is that it has an extremely small standard deviation.

Finally, method 2 is nearly the same as method 3 which is expected as the two distributions are somewhat similar. But notice that the fraction of values z larger than y is extremely high, although its standard deviation is much lower compared to method 3. Hence the method does a bit weaker than method 3. In summary, it is clear

Table 6.3
Test data results for using 5 methods in Elia Data

Metric	Method 1	Method 2	Method 3	Method 4	Method 5
Pinball loss	27.15	25.45	24.69	30.49	21.42
$\mathbb{P}\{z > y\}$	571/907	902/907	888/907	888/907	835/907
$stdev(z)$	264.98	306.94	328.27	459.23	408.24

that all these methods have their ways to have higher or lower values as seen above. Now, next we approach a case where we consider multiple values of τ, not just one.

SAMPLES FOR VARIOUS τ VALUES

Essentially, z is equal to $F^{-1}(\tau)$ when we know it (but usually we do not and have a guess for it). So, for a moment let us call z as z^τ as the value of z greatly depends on τ. When we have only one value of τ for the whole set, it does not matter. But here we will consider multiple values of τ. In particular, we develop the pinball loss function, i.e., the function value when CDF $F(y)$ which is unknown is forecasted for 9 different values $F(P_{10})$, $F(P_{20})$, ..., $F(P_{90})$. We could also do the same with 19 different functions (every 5 percentile points), as the analysis is similar.

So, if an algorithm makes n forecasts y_1, y_2, ..., y_n, but before each forecast (or more forecasts) let the predictions be z_1^τ, ..., z_n^τ, for each τ in 0.1, 0.2, ..., 0.9, then the statistic (for some weights w_τ) to compare is

$$\sum_{\tau=0.1}^{0.9} w_\tau \sum_{i=1}^{n} L_\tau(y_i, z_i^\tau)$$

where w_τ could be simply $1/n$ if one wishes to have all the τ to have equal weights (which is what we will do throughout this document). In fact even $1/n$ is not needed as from an individual standpoint it does not change. If there are multiple forecasting algorithms, the winning algorithm is determined by the one with lowest value of the above statistic.

We consider 3 months of 1-hour data from Golden, CO for this exercise. The months covered are September to November 2017. We use unscaled data for the analysis. We make a prediction 24-hours in advance. Hence we consider two lags, one is 1 day in advance and another is 2 days in advance for the same hour. Besides that, we also use hour, the forecasted temperature, and Denver weather predictions to predict the column called global which is the irradiance measured. The analysis naturally requires us to consider special weather types as in Section 4.4. In fact the only difference between the results there and here is that here we consider the unscaled global irradiance.

In fact we only consider the first month's data after removing the first two days which would have resulted in NaNs with respect to the lag features. Then we train our models using the first 2 weeks of data and test them with the next 2 weeks without retraining along the way (except we do include the most appropriate lag features). Thus, from a model standpoint they are included but the model itself is not retrained. Then we create a point forecast for the test data using an XGBoost regressor with with objective as "absolulteerror", `n_estimators = 50`, `max_depth = 5`, `learning_rate = 0.1`, and `gamma = 5.0`. It is important to notice that these have not been selected to be optimal, and other metrics have been asked to be at the default values.

The point forecast that we assume using the XGBoost regressor is the mean. Now, we fit the variance. We consider two models for variance. For the first model we assume that all the time intervals have the same variance, i.e., they are homoscedastic. Then we simply estimate the variance as the average sum of squares in the training data between the actual values and the forecasted values which is the mean. Then for the test data variance we simply use the same values as in the train data. But for the mean, we compute the test data's XGBoost point forecast using the model trained. For the second model, we use variance corresponding to each hour of the day and use that, while using the same model for the mean.

We consider two different approaches to compute z^{τ}, but these are the same as the previous example, namely the Gaussian and the gamma approaches. For the Gaussian, for a given τ, we compute z^{τ} as the inverse of the normal CDF with mean equal to the mean computed using XGBoost regressor and the standard deviation using the two variance approaches, one using a common for all values of hour, while the other is hour-specific. The inverse of the CDF is computed at τ. For the gamma distribution z^{τ} is the inverse of the CDF of a gamma random variable with given scale and shape parameters which are shown to be computed.

The shape and scale parameters are related to the mean and variance using the fact that the scale parameter is just the variance divided by the mean, and the shape parameter is simply the mean divided by the scale parameter. Hence using these scale and shape parameters we can compute the inverse at τ. For both Gaussian and gamma distributions we use

$$\sum_{\tau=0.1}^{0.9} \frac{1}{n} \sum_{i=1}^{n} L_{\tau}(y_i, z_i^{\tau})$$

to compute the pinball loss functions. The results in Table 6.4 are almost equivalent across the two methods and across the way we would compute standard deviation.

Table 6.4

Test data results for using 2 methods in Golden, CO September 2017 Data

	Gaussian approach	Gamma Approach
Pinball loss Constant standard deviation	120.8016	119.2242
Pinball loss Varying standard deviation	120.5171	118.9254

If one were to become extremely precise, then the Gamma distribution with hourly varying standard deviation produces the best results.

6.3 PARAMETRIC METHODS

We have seen two models thus far in the previous section using Gaussian distribution and gamma distribution. Also, in one case we used the mean as the day-ahead forecast while in the other case we had used an XGBoost model to obtain the point forecast. In both Gaussian as well as gamma distributions, irrespective of how we got the point forecasts, in both cases we had used the standard deviations. Again, in one case we had assumed a constant standard deviation but in the other case we used a time-wise constant standard deviation. We will continue that from in this section as well by using XGBoost for the point forecast.

However, for the standard deviation, we assume they remain a constant throughout. For that, we consider a special case of scaled forecast that results in values being about the same each time unit when there is sun. However, it also necessitates us to consider two other distributions for the total scaled power, namely, beta distribution and exponential distribution. They are mainly to show that other distributions are also possible. Further we have just used Gaussian and gamma. Also, we first show the results using simulated data and slowly build the case for real data. In the process we will explain how this works.

Here we consider parametric methods for distributional solar forecasts. Let S_n be the scaled solar irradiance at the nth hourly global horizontal observation. We consider a special case where we are able to divide the solar irradiance by the maximum possible value, and hence the observed S_n is in $[0,1]$ as it is the observed irradiance over the maximum possible during that month. How does the marginal distribution of S_n look like? We plot those across the three data sets we have seen in Section 4.4 but the values are scaled from 0 to 1. Although we show for two distributions, we will end up using only for the case of beta.

In Figure 6.1 we plot the distributions of the scaled forecasts as well as those of (1-scaled) forecasts at the three locations that we previously considered. In particular, one is SolarTAC which is based out of Aurora, CO. The second is based out of Golden, CO and it is called SRRL. The third is from Los Angeles, CA which we call LAX. Besides plotting the histogram, we also plot distributions (where the means and standard deviations fit) of beta and exponential distributions. The actual fits are in red color displayed on the histograms. Although they are not the greatest fits, they are reasonable.

SIMULATED SAMPLING FROM A BETA DISTRIBUTION

To help understand how the model works, we first demonstrate for a simulated experiment where we know the values the resulting variable should take. Recall from the previous chapter that when the underlying data is not observable, it is not possible to

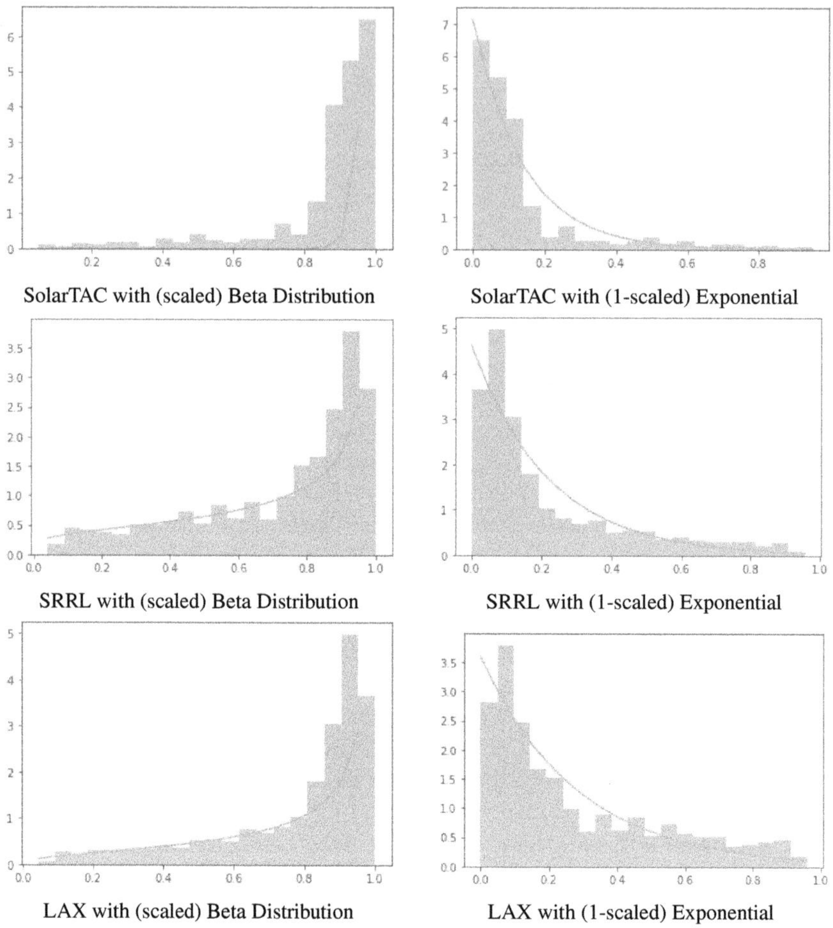

Figure 6.1 Marginal Distributions of Scaled Irradiance Fitted Against Beta and Exponential

compare anything with the ground truth. So we take a simulated approach initially. We will eventually add some noise and finally consider solving the data in Figure 6.1. We take the simple problem of forecasting one time period in advance. We also try and select our mean to vary across time while keeping our standard deviation a constant.

In that light, suppose that we know S_n, the nth scaled irradiance. We now predict S_{n+1} given S_n. In particular, we generate S_{n+1} as sampled from a random variable according to a Beta distribution with mean $\frac{S_n+0.5}{2}$ and standard deviation 0.25 (notice that would ensure that S_n is within the range $[0,1]$ and also that the random variable itself ensures that). The mean is guaranteed to be between 0.25 and 0.75 with standard deviation always at 0.25 (hence it is satisfies our example of having a mean that varies but the standard deviation remains fixed).

We compute the 70th percentile or P_{70} using the inverse CDF of S_{n+1} at 0.7, i.e.,

$$\mathbb{P}\{S_{n+1} \leq P_{70}\} = 0.7$$

which is straightforward to compute if we know S_n as it is nothing but the inverse of the CDF at 0.7, since we know the mean and standard deviation. Although in practice P_{70} is not observed, in the simulation it is known, hence we see how close to it our simulations are. But for the simulation as well as to predict P_{70}, we need a few relations in terms of the parameters α and β.

Note that a Beta random variable with support $[0,1]$, and shape parameters α and β has a mean $\alpha/(\alpha+\beta)$ and variance $\frac{\alpha\beta}{(\alpha+\beta)^2(\alpha+\beta+1)}$. In our example, we are given the mean as μ and variance as σ^2, and using them we need to estimate α and β. From the method of moments we can estimate the values of α as

$$\alpha = \left(\frac{\mu(1-\mu)}{\sigma^2} - 1\right)\mu$$

and β as

$$\beta = \left(\frac{\mu(1-\mu)}{\sigma^2} - 1\right)(1-\mu)$$

from which we could obtain μ and σ^2. Now, using α and β, we can generate random numbers according to the Beta distribution.

Now we will see how generating simulated data for verification works. We initialize by letting $\mu_0 = 0.5$ and $\sigma_0 = 0.25$. Then, for $n = 0$ to N (for some arbitrarily set quantity N), we compute $\mu_{n+1} = \frac{S_n + 0.5}{2}$ and $\sigma_{n+1} = 0.25$. Then we obtain the parameters

$$\alpha_n = \left(\frac{\mu_n(1-\mu_n)}{\sigma_n^2} - 1\right)\mu_n$$

and

$$\beta_n = \left(\frac{\mu_n(1-\mu_n)}{\sigma_n^2} - 1\right)(1-\mu_n)$$

which correspond to the mean and variance described before. Of course all these assume we know S_n. We randomly generate S_n by sampling from a Beta distribution with parameter α_n and β_n for all $n \geq 0$. We obtain $P_{70,n}$ by computing the inverse of the CDF of a Beta random variable with parameter α_n and β_n, where $P_{70,n}$ is the nth value of P_{70}.

We use a SARIMAX model for our analysis. Of course, the only input we use in our analysis is the time series $S_0, S_1, \ldots, S_N$. Everything that we were presented is only used to verify our approach. In our results, we set aside some of the data for training and use a sliding window for the remainder (that means we retrain every time unit). We use a SARIMAX model with no seasonality and no exogenous variables. As expected, an AR(1) model produced the lowest Akaike's Information Criterion (AIC) since we know from the simulated data (but not in this experimental setting), the actual values depend only on the most recent observations. Figure 6.2 provides the autocorrelation function for the data.

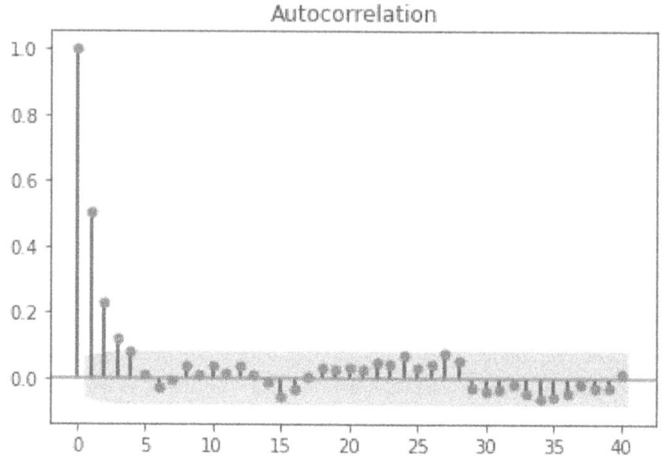

Figure 6.2 Autocorrelation of the Data for SARIMAX Calculations

What auto-ARIMA for SARIMAX gives us are two things: an estimate of S_n that we call $\hat{S}_n$ and a confidence interval which is $(\hat{S}_n - 1.96\hat{\sigma}_n^{\varepsilon}, \hat{S}_n + 1.96\hat{\sigma}_n^{\varepsilon})$ where $\hat{\sigma}_n^{\varepsilon}$ is an estimate of the standard deviation of the residuals. For our parametric model we assume that S_n is according to a Beta distribution with mean $\hat{S}_n$ and standard deviation $\hat{\sigma}_n^{\varepsilon}$. Note that this is our only guess based on the autocorrelation given in Figure 6.2 as there is no way for us to know the exact relationship between the mean and S_n.

Thus, for all n in the test set, $\hat{\mu}_n = \hat{S}_n$ and $\hat{\sigma}_n = \hat{\sigma}_n^{\varepsilon}$. Then we obtain the parameter estimates as

$$\hat{\alpha}_n = \left(\frac{\hat{\mu}_n(1-\hat{\mu}_n)}{\hat{\sigma}_n^2} - 1 \right) \hat{\mu}_n$$

for $\hat{\alpha}_n$ and

$$\hat{\beta}_n = \left(\frac{\hat{\mu}_n(1-\hat{\mu}_n)}{\hat{\sigma}_n^2} - 1 \right) (1-\hat{\mu}_n)$$

for $\hat{\beta}_n$. We assume that S_n is randomly generated by sampling from a Beta distribution with parameter $\hat{\alpha}_n$ and $\hat{\beta}_n$. Using that, we obtain $\hat{P}_{70,n}$ by computing the inverse of the CDF of a Beta random variable with parameter $\hat{\alpha}_n$ and $\hat{\beta}_n$ For our test samples, $\hat{P}_{70,n}$ was within 2% of $P_{70,n}$ on average.

ADDING NOISE TO SIMULATED DATA

This is a very similar experiment but with a small change to add noise to the simulated data. So we simply present the setting and discuss the various steps involved. We initialize by letting $\mu_0 = 0.5$ and $\sigma_0 = 0.25$. Then, for $n = 0$ to N, we obtain the parameters

$$\alpha_n = \left(\frac{\mu_n(1-\mu_n)}{\sigma_n^2} - 1 \right) \mu_n$$

and

$$\beta_n = \left(\frac{\mu_n(1-\mu_n)}{\sigma_n^2} - 1\right)(1-\mu_n)$$

with the only difference being how we obtain μ_n and σ_n. But once we have our S_n characterized, we randomly generate S_n by sampling from a Beta distribution with parameter α_n and β_n.

We obtain $P_{70,n}$ by computing the inverse of the CDF of a Beta random variable with parameter α_n and β_n. Then we generate uniform $(0,1)$ random variables U_n and U_n' that are both independent. The key difference (and thus all the others are different too) is that we compute $\mu_{n+1} = \frac{S_n+U_n}{2}$ and $\sigma_{n+1} = U_n'/5$. We next estimate the quantiles for noise-added simulation. For all n in the test set, $\hat{\mu}_n = \hat{S}_n$ and $\hat{\sigma}_n = \hat{\sigma}_n^{\varepsilon}$. However, because of our noise structure $(\hat{\sigma}_n^{\varepsilon})^2$ will actually estimate $(\mathbb{E}[U_n'/5])^2 + \mathbb{V}[U_n/2] = 0.01 + 1/48$.

The second term above is additional noise while the first term is what is used in sampling from Beta in simulations. The remaining steps are the same, i.e., obtain $\hat{\alpha}_n$, $\hat{\beta}_n$, and obtain $\hat{P}_{70,n}$ by computing the inverse of the CDF of a Beta random variable with parameter $\hat{\alpha}_n$ and $\hat{\beta}_n$. For our test samples, $\hat{P}_{70,n}$ was within 13.7% of $P_{70,n}$ on average.

USING SOLAR IRRADIANCE DATA

In this portion, we use day ahead hourly data from SolarTAC (from Aurora, CO), SRRL (from Golden, CO), and LAX (from Los Angeles, CA) as before to obtain scaled solar irradiance. There are some differences, however. First of all, we only consider 28 days of data. Also, we consider the first year for training, and use the second year for testing (without retraining) the model. Of course, we do use the features a day ahead. So we essentially use the previously trained model but use the most appropriate features for the model. Since this is a day-ahead model, we use the features from the previous day for the model.

Suppose that we have the time series S_0, S_1, ..., S_N which are the scaled values. The key difference here is that the scale reduces all the values to be within 0 to 1. For the SARIMAX and Exponential Smoothing models we do not use any exogenous variables, but for the XGBoost and the gradient boosting models we use data regarding the temperature forecasts and the weather forecasts. We also consider the day ahead, 2 days ahead, and the day ahead plus one solar data for the XGBoost and gradient boosting approaches. We also use the hour data. Of course, the date-time feature would be removed before all the experiments.

In terms of the modeling parameters, we used a set without trying to optimize the features. We also kept the values the same across the three regions. For XGBoost model we used `n_estimators=400`, `max_depth=8`, `learning_rate =0.01`, `objective` as 'reg:absoluteerror', `subsample=1.0`, `min_child_weight=25`, and `gamma=20.0`. The rest used the default values. Likewise for the gradient boosting methods we used `n_estimators=100`, `max_depth=5`, and `min_samples_leaf=3`.

All the other variables were kept as the default values. The value can certainly be improved by selecting other values.

For the exponential smoothing model, we used additive trend and multiplicative seasonality, like we did earlier. Then we predict 8 horizons in advance and take the last one as we assume we need to predict a whole day in advance. Likewise, for the SARIMAX approach, we use $(p,d,q) = (2,1,0)$ and (P,D,Q) as $(0,0,1)$ with 8 seasonal periods. We did not try different values and let things be the same for all three regions. Although in practice we would try to tune as accurate model as possible for a given location. We may also want to consider including the exogenous features. Also, we used a persistent forecast where the assumption was that we would forecast the value observed a day in advance.

We consider the month of October for both SolarTAC and SRRL data, while we use the middle of November through the middle of December for LAX. For those reasons we only consider the 8 hours of reasonable solar times. We use 28 days of data from one year to train and use the next year same dates for testing. In all the methods we had a choice of dividing the day into either hours and provide hourly standard deviations. However, we chose to not go that route and simply used a constant value for all the data points. In future it may be a better idea to simply use hourly forecasts of variance.

For all 5 methods: exponential smoothing, SARIMAX, XGBoost, Gradient Boosting, and persistent methods, we obtained the mean as what the methods obtained. Thus the variance was also different since the means were different. For the variance we used the mean of *(actual - forecasted)* values for all the methods. Since the forecasted values are different for each method, as a result the variance was also different. For all the methods we did not use an hourly forecast of variance, although that was an option indeed. For all $n = 1$ to N, we obtain an estimate of S_n that we call $\hat{S}_n$ and a standard deviation that we call $\hat{\sigma}_n^{\varepsilon}$.

For our parametric model we assume that S_n is according to a Beta distribution with mean $\hat{S}_n$ and standard deviation $\hat{\sigma}_n^{\varepsilon}$. For all n in the test set, $\hat{\mu}_n = \hat{S}_n$ and $\hat{\sigma}_n = \hat{\sigma}_n^{\varepsilon}$. We use that to obtain the parameter estimates $\hat{\alpha}_n$ and $\hat{\beta}_n$ which are needed for the Beta distributions. We assume that S_n is randomly generated by sampling from a Beta distribution with parameter $\hat{\alpha}_n$ and $\hat{\beta}_n$. We obtain the pinball loss function for all values of q from 0.1 to 0.9 and simply take the averages.

In all three experiments (see Table 6.5), the best method in the set was SARIMAX with exponential smoothing performing only slightly worse. It is possible that either with better tuning or more accurate data, the XGBoost and the gradient boosting approaches could be much better. But they are all quite comparable. In comparison, the persistent method performed much worse. In all three data sets, for the exponential smoothing and the SARIMAX, the MAPE was about 22-32% which is quite high. Interestingly, although the MAPE was much higher for XGBoost and gradient boosting, it had much smaller median percentage error. That may be because of the objective selected in the machine learning training.

Although we only present the case of scaled data with beta distribution, we have already seen unscaled data in the previous section and model it using Gaussian and

Table 6.5
Pinball loss average for using 5 methods in Golden CO, Aurora CO, and Los Angeles CA

Method	SolarTAC Data	SRRL Data	LAX Data
Exponential Smoothing	0.0533	0.0693	0.0655
SARIMAX	0.0527	0.0675	0.0626
XGBoost	0.0837	0.0936	0.0928
Gradient Boosting	0.0769	0.0936	0.1086
Persistent	0.1177	0.1365	0.1128

gamma distributions. However, it is important to iterate that it would be worthwhile to try out all the distributions that would be choices based on the histograms. Here in all cases we obtained two statistics: mean and variance. But some distributions may require more than two. Also, it is not necessary to use method of moments but the maximum likelihood estimate might be more appropriate in some settings. Also, we have tried a few algorithms such as SARIMAX, exponential smoothing, XGBoost, and Gradient Boosting. But one could consider other methods as well.

6.4 NON-PARAMETRIC METHODS

Unlike parametric methods where we fit a distribution for the data, here we do not make any such assumptions. We simply use the pinball loss function to determine the best fit. However, we take our time to explain it step by step. For the non-parametric methods for distributional solar forecasts, we let S_n be the scaled solar irradiance at the nth hourly global horizontal observation. There is really nothing about it being hourly, the same method would also work with every 15 minutes or every 5 minutes. It does make it easier by sticking to a notation. But we do not necessarily make the assumption that S_n is in $[0,1]$, in fact we do not even need it to be scaled.

The usual forecasting techniques that we have seen in previous chapters essentially estimate

$$\mathbb{E}[S_{n+k}|S_0, S_1, \ldots, S_n, \mathbf{V}_{n+k}]$$

for some $k > 0$, any n in test set, and vector of exogenous variables $\mathbf{V}_{n+k}$. What we need for distributional forecasts is to estimate y for some target value of

$$\mathbb{P}\{S_{n+k} \leq y|S_0, S_1, \ldots, S_n, \mathbf{V}_{n+k}\}$$

for some $k > 0$, n in test set, and exogenous variables $\mathbf{V}_{n+k}$. We are usually interested in estimating quantiles of y such as P_{10}, P_{20}, …, P_{90}, also called deciles.

We will show the non-parametric methods for a specific target quantile τ (for example, $\tau = 0.7$ would estimate P_{70}) but with the understanding that we will then extend it to the case of all values of τ that is of interest (such as $0.1, 0.2, \ldots, 0.9$). For this we consider a quantile regression model where Y is a scalar outcome variable and $\mathbf{X}$ is a vector of regressors. In the usual regression, we obtain the relationship between $\mathbf{X}$ and Y through the mean, i.e., an estimate of $\mathbb{E}[Y|\mathbf{X}]$. However, in a quantile regression, we obtain the relationship between $\mathbf{X}$ and Y through the quantiles, i.e., an estimate of $y(\tau)$ so that $\mathbb{P}\{Y \leq y(\tau)|\mathbf{X}\} = \tau$.

The reasons for taking the regression route are: ease of explanation, and availability of appropriate Python libraries. But we will subsequently switch to XGBoost. We will subsequently see that the only real change is the cost function. Recall that the usual forecasting techniques essentially estimate

$$\mathbb{E}[S_{n+k}|S_0, S_1, \ldots, S_n, \mathbf{V}_{n+k}]$$

for some $k > 0$, any n in test set, and vector of exogenous variables $\mathbf{V}_{n+k}$. What we need for distributional forecasts is to estimate y for which

$$\mathbb{P}\{S_{n+k} \leq y|S_0, S_1, \ldots, S_n, \mathbf{V}_{n+k}\} = \tau$$

for some $k > 0$, n in test set, and exogenous variables $\mathbf{V}_{n+k}$.

We first show how to predict the median using quantile regression, so P_{50} in this case. Later we would generalize to any τ using a method we have seen before. Say that we have N training samples for a regression. Let us denote the samples as y_i and $\mathbf{X}_i$ for all i in $[1,N]$. Assume that the very first element of every $\mathbf{X}_i$ is 1 (standing for the intercept) so that we can write the expressions crisply. In the usual regression, the optimal coefficient vector $\hat{\beta}$ is obtained that minimizes

$$\sum_{i=1}^{N}(y_i - \mathbf{X}_i^T\beta)^2$$

and is typically written as $\hat{\beta} = (\mathbf{X}^T\mathbf{X})^{-1}\mathbf{X}^T y$.

In quantile regression for median, the optimal coefficient vector $\hat{\beta}_{0.5}$ is obtained that minimizes

$$\sum_{i=1}^{N}|y_i - \mathbf{X}_i^T\beta|$$

which is not easy to write down since it is not differentiable everywhere. At any rate, the median regression is also known as LAD (least absolute deviation) regression. Median regression is more robust to outliers (just like how median is in comparison to the mean). However, since the absolute value function is not differentiable everywhere, one uses linear programming techniques to solve, although that is not shown here.

But we can show that in median regression, the optimal coefficient vector $\hat{\beta}_{0.5}$ is obtained that minimizes

$$\sum_{i=1}^{N}|y_i - \mathbf{X}_i^T\beta|.$$

For that we recall the pinball loss function

$$L_\tau(y,z) = \begin{cases} (y-z)\tau & \text{if } y > z \\ (z-y)(1-\tau) & \text{if } y \leq z. \end{cases}$$

For the median $\tau = 0.5$ and we can rewrite

$$L_{0.5}(y,z) = 0.5|y - z|$$

using the above expression. We established in Section 6.2 that the value of z that minimizes

$$\mathbb{E}[L_{0.5}(Y,z)]$$

is $F^{-1}(0.5)$ which is the median. Thus, in the test sample, we estimate the median when we are given the jth sample $\mathbf{X}_j$ as $\mathbf{X}_j^T \hat{\beta}_{0.5}$.

The median analysis can be easily extended to generic target quantiles. For that, we consider a generic quantile target τ. In quantile regression, the optimal coefficient vector $\hat{\beta}_\tau$ is obtained that minimizes

$$\sum_{i=1}^{N} L_\tau(y_i, \mathbf{X}_i^T \beta)$$

where

$$L_\tau(y,z) = \begin{cases} (y-z)\tau & \text{if } y > z \\ (z-y)(1-\tau) & \text{if } y \leq z. \end{cases}$$

We established in Section 6.2 that the value of z that minimizes

$$\mathbb{E}[L_\tau(Y,z)]$$

is $F^{-1}(\tau)$. Thus, in the test sample, we estimate the quantile corresponding to target τ when we are given the jth sample $\mathbf{X}_j$ as $\mathbf{X}_j^T \hat{\beta}_\tau$.

The minimization is usually done using linear programming techniques. But quantile regression is available in statistical software packages already have these built in. Thus, if we were using multiple linear regression, the main step is to use the pinball loss function as the cost function to be minimized is (with unknowns β)

$$\sum_{i=1}^{N} L_\tau(y_i, \mathbf{X}_i^T \beta)$$

so the optimal value is $\hat{\beta}$. So, when we write, the β terms need to be handled carefully as

$$L_\tau(y_i, \mathbf{X}_i^T \beta) = \begin{cases} (y_i - \mathbf{X}_i^T \beta)\tau & \text{if } y_i > \mathbf{X}_i^T \beta \\ (\mathbf{X}_i^T \beta - y_i)(1-\tau) & \text{if } y_i \leq \mathbf{X}_i^T \beta \end{cases}$$

where the known is $\hat{\beta}$.

Even while using XGBoost instead of multivariate regression, the approaches use gradient based steepest descent methods in their code which require taking partial derivatives. Since the pinball loss function is not differentiable everywhere, we would have to solve it as a constrained optimization problem. Assuming that is not an issue, before we use XGBoost in our regression, we wish to showcase the SARIMAX method. Recall that the default is sum-of-squared-errors (SSE) in all our forecasting models using SARIMAX. This simplistic approach leverages that and so we use SSE as usual for our forecasting methods to estimate

$$\hat{S}_{n+k} = \mathbb{E}[S_{n+k}|S_0, S_1, \ldots, S_n, \mathbf{V}_{n+k}]$$

for some $k > 0$, any n in test set, and vector of exogenous variables $\mathbf{V}_{n+k}$.

We use a cross-validation dataset where we have actual values of S_{n+k} (besides the forecasted value). The simplistic approach is to develop a quantile regression with S_{n+k} as the outcome variable (i.e., y) and $\hat{S}_{n+k}$ as the regressor (i.e., $\mathbf{X}$). The key idea is to showcase how the quantile regression works (but not to suggest how it must be done for forecasting). Further, it is also used to compare the results obtained using the beta distributions in the forecasts in the previous section. In general 70% of our P_{70} predictions must be smaller than P_{70}, and the rest higher.

We first estimate the quantiles for simulated data, like we did in the previous section. We use the same simulated data as before where we know the ground truth of the quantiles. We obtain forecasts using SARIMAX first. We do not assume any distribution and simply use quantile regression between the observed value and the forecasts. The P_{70} regression yielded coefficients of 0.1704 and 1.0036 for the intercept and the slope of the fitted regression line, respectively. For our test samples, $\hat{P}_{70,n}$ was within 2.9% of $P_{70,n}$ on average. It is important to notice that it would not be appropriate to compare against the values used in the previous section.

If we were to use solar irradiance data where we consider day ahead hourly data from SRRL (Golden, CO) as before to obtain scaled solar irradiance. We take a similar approach as the simulations in SARIMAX except in this case we do not know the ground truth values. Nonetheless, we use the time series S_0, S_1, ..., S_N. We set aside the first 61 days for training and use the remaining for testing using a sliding window. We use XGBoost for our forecast to be used in the quantile regression. Then using the XGBoost forecast as regressor and the actual solar scaled irradiance as observed value, we perform quantile regression. We obtain the deciles from 0.1 to 0.9 and for each decile we compute the intercept and slope of the fitted line. The results are shown in Figure 6.3.

It is worthwhile explaining Figure 6.3 briefly. The line in orange or thick black is the ordinary least squares (OLS) line which is a regression between the XGBoost predictions and the actual values using simple linear regression. Notice that although the OLS line is aligned with the other lines, its slope is a little different. The other black colored doted lines are the likeness corresponding to P_{10} (bottom) to P_{90} (top). Notice that the OLS is fairly close to the P_{50} line. Also notice that a fraction τ of the points are below the line corresponding to $P_{100\tau}$. But what we are interested is not in a regression between the XGBoost prediction and the actual value, but a quantile forecast.

QUANTILE FORECAST

Now we are ready to use XGBoost and nothing else for quantile forecast. We present three use cases. Each time we obtain the pinball loss functions for the three cases. We compare four methods: XGBoost quantile regression, conformal predictions,

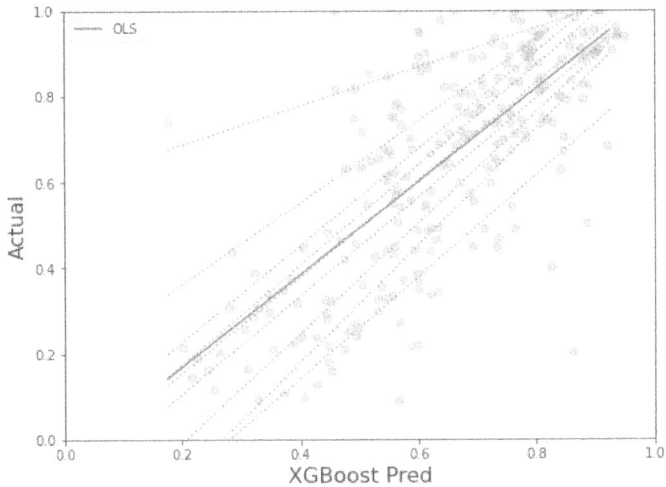

Figure 6.3 Quantile Regression and Actual Values with XGBoost

Gaussian quantile regression, and a naive approach. The first two are based on non-parametric quantile forecast. We only use Gaussian quantiles for the third approach which is indeed a parametric approach. Finally, as a way to compare the approaches we also consider a naive approach that we will explain. In all four approaches, we use the same features and same train/test data. We also use XGBoost modeling for the forecast.

We have seen various instances of all the three test cases in this book. So until the time the cases are presented, we will leave them out. We will focus (out of the four) on the two approaches we have not seen, in some detail, i.e., the XGBoost quantile regression and conformal predictions. The main difference between using those versus a parametric approach such as Gaussian is that in parametric methods, it combines bias and variance, so it's not purely a measure of the standard deviation of errors, unless bias is negligible. In addition, the residuals need to be homoscadastic with the appropriate distribution (Gaussian in this case).

In comparison, quantile regression with XGBoost can directly use the objective as "reg:quantileerror" which is essentially the pinball loss function. This gives direct interval estimates (e.g., 10th and 90th percentiles). Further, conformal predictions are a distribution-free, powerful method that gives valid intervals under mild assumptions. It would be possible to train a model (such as XGBoost), and then use a calibration set to estimate residuals and derive quantiles. It works regardless of the underlying model. There are other approaches such as Bayesian approaches and ensembling with variance estimation that are not considered here.

Before presenting the python code for the cells corresponding to the XGBoost quantile regression and the conformal predictions, we describe the first of the three case studies. We consider a much larger example of Elia [6] where we consider data from across 8 years (from 2012 to 2019) and the data being from October to December of each year. We train our model using years 2012 to 2018, and we use 2019 to test. However, recall from the description in the previous chapter that this data is from multiple cities and year after year more solar PV cells are being added. There is a column called solar capacity which indicates the amount of solar available.

So for each year we divide the "solar power" by the "solar capacity" so that we get roughly the power corresponding to a single unit. This we call as the "scaled solar". Therefore, instead of predicting the solar power, our models predict the "scaled solar" which we hope is about similar from year to year. We also only use 7 hours of day time for the data, which is from 9am to 3:45pm in 15-minute intervals. Hence we have 28 data points for each day. We use a lag of 28, 29, 30, and 31 points (which corresponds to the data observed the previous day's four 15-minute intervals). In addition, we also use the "hour" of the day as a feature.

In addition to these features, we also merge with the weather forecast in the town on Antwerp in Belgium. Understandably, all the data points are not in Antwerp but there are a good number of them. Also, the country of Belgium is small enough and we hope this would suffice. Thus in summary, we have the temperature, weather category (which has been simplified into three: overcast, scattered, clear), wind speed, humidity, and barometer. Thus, to predict the "scaled solar", we use features such as temperature, weather category, wind speed, humidity, barometer, hour, lag 28, lag 29, lag 30, and lag 31.

Now we are in a position to discuss a cell in the python code dedicated to the XGBoost quantiles and the conformal prediction methods for the Elia data. This is described in Figure 6.4. Line 2 essentially states the quantile values 0.1, 0.2, ..., 0.9 using numpy arrange. Line 3 initializes the vector corresponding to the XGBoost quantile predictions as an empty vector. Lines 5 through 12 describe the process of obtaining these quantiles for the XGBoost. Line 5 essentially iterates through the 9 values of the quantiles (0.1 to 0.9). Lines 6 through 8 describe the XGBoost model used. This has not been optimized and can further be fine tuned.

The most important is the objective. The reason it says "reg:quantileerror" is because it directly uses the pinball loss function at that quantile q. That is exactly the next item, `quantile_alpha`. The other items can be changed and these have been used elsewhere in the book, hence is not described here. Then in line 9 we train the model using the 7 years of information we have. In line 10, we obtain the scaled predictions for the test data using just the test data. Then in line 11, these values are returned to the original ones by multiplying by "solar capacity". Thus in line 12, we have the quantile predictions for quantile q.

The remaining lines correspond to the second method, namely conformal predictions. In lines 15-17 we describe the point forecast model first. Here notice that this is a standard XGBoost approach for point forecast. For example, the objective is "reg:absoluteerror" which is what is used if we were to obtain the smallest absolute error in the training data. All other parameters are exactly the same as the previous case. Then the point forecast model is fit in line 18. Then in line 19, we predict the *train* data to see what training sample will be obtained. Then it is scaled, like we did

```
[4]:
1 # --- Quantile Regression with XGBoost ---
2 quantiles = np.arange(0.1, 1.0, 0.1)
3 xgb_quantile_preds = {}
4
5 for q in quantiles:
6     q_model = XGBRegressor(
7         objective='reg:quantileerror', quantile_alpha=q, n_estimators=50, max_depth=5,
8         learning_rate=0.1, alpha=6.0, min_child_weight=5.0, gamma=30.0, random_state=42)
9     q_model.fit(X_train, y_train)
10    preds_scaled = q_model.predict(X_test)
11    preds = preds_scaled * test['Solar Capacity']
12    xgb_quantile_preds[q] = preds
13
14 # --- Conformal Prediction ---
15 point_model = XGBRegressor(
16     objective='reg:absoluteerror', n_estimators=50, max_depth=5, learning_rate=0.1,
17     alpha=6.0, min_child_weight=5.0, gamma=30.0, random_state=42)
18 point_model.fit(X_train, y_train)
19 train_point_pred_scaled = point_model.predict(X_train)
20 train_point_pred = train_point_pred_scaled * train['Solar Capacity']
21 residuals = train['Solar Power'] - train_point_pred
22
23 conformal_quantile_preds = {}
24 for q in quantiles:
25     offset = np.quantile(residuals, q)
26     test_point_pred_scaled = point_model.predict(X_test)
27     test_point_pred = test_point_pred_scaled * test['Solar Capacity']
28     conformal_quantile_preds[q] = test_point_pred + offset
```

Figure 6.4 Training cell in quantile regression and Conformal prediction

earlier, by multiplying by "solar capacity".

Then the critical part of this model emerges. In line 21, we obtain the residuals, which is the difference between the trained "solar power" and what is predicted. There are used in lines 23-28 to obtained the conformal predictions which is initialized as an empty vector in line 23. Then for each of the 9 deciles (0.1 to 0.9) in line 24 we perform separate analysis. We derive the offset using the residuals and quantile q taking the values in the quantile function of numpy in line 25. Then line 26 gives the point forecast and line 27 scales it by the "solar capacity". Then the conformal quantile predictions for quantile q is given as the sum of the point forecast and the offset.

Next, to assess how the quantile forecasts were, we describe them in Figure 6.5. In lines 2 to 4, we describe the pinball loss function that is used not only for these two methods but the other two as well (Gaussian and the naive methods). Given a quantile q, the pinball loss function computes

$$\frac{1}{n}\sum_{i=1}^{n} L_\tau(y_i, z_i^\tau)$$

assuming there are n sample values in the test set. Notice that the values `y_true` and `y_pred` correspond to the values of y_i and z_i vectors for all i.

In lines 6 to 14 we provide the method name and pinball loss values to the `results` that is initialized in line 6 to be an empty vector. Then, in lines 7 and 8, we consider the two methods and the quantile predictions for each q. Although

```
[5]:
1 # --- Pinball Loss ---
2 def pinball_loss(y_true, y_pred, q):
3     diff = y_true - y_pred
4     return np.mean(np.where(diff >= 0, q * diff, (1 - q) * (-diff)))
5
6 results = []
7 for method_name, preds_dict in [('XGB_Quantile', xgb_quantile_preds),
8                                 ('Conformal', conformal_quantile_preds)]:
9     losses = []
10    for q, preds in preds_dict.items():
11        loss = pinball_loss(y_test_actual.values, preds.values if hasattr(preds, 'values') else preds, q)
12        losses.append(loss)
13    avg_loss = np.mean(losses)
14    results.append((method_name, avg_loss))
15
16 # Compare methods
17 for method, loss in results:
18    print(f"{method} Average Pinball Loss: {loss:.4f}")
```

Figure 6.5 Evaluating test using pinball in quantile regression and Conformal prediction

lines 10-14 look complex, all it does is to sum up the pinball loss for the given value of q. Most of that is done in line 11 while the other lines essentially are there for book keeping. Finally the averages are computed in line 13. Then in line 14 we obtain the results for the two sets. Lines 17 and 18 are just there to output the results.

Afterward, we also consider the two methods namely Gaussian forecast and naive approach. The Gaussian forecast is similar to what we have seen in the previous section, where we use XGBoost to predict the point forecast (in fact we simply use the output from conformal predictions). We also compute the standard deviation by taking the difference between the point forecast and the actual values in the train set. Then we use the mean and standard deviations as the parameters of the Gaussian distribution and obtain the predictions for the test data. Then we compute the pinball loss functions for them.

For the naive time of the day quantiles, we group the training data each hour. Then within each hour we obtain the mean values. Then we merge the mean values to the testing data. Then for each day we predict the same values as the mean values. This will give us an idea of how the values are if we only used the mean values and not individual values. Then we obtain the pinball loss functions by comparing the forecast against the actual values. We describe the results for two other test cases namely the SRRL 2017 scaled solar hourly data and the SolarTAC 15-minute unscaled solar data as well.

Although we provide the results in a single table, the three data sets are extremely different. Before presenting the results, we give some more details of the experiments for the two CO data sets. For the SRRL data, we only consider 1 hour granularity and we scale the results. We consider features such as 1 seasonal period ahead, 1 seasonal period plus 1 hour ahead, and 2 seasonal period ahead. In addition, we use the Denver weather as features, that includes Denver temperature as well as some of the major weather conditions. We also include hour as a feature. We take the first seven days as training and the remaining nearly 3 months as testing. We do not retrain the model but use the appropriate day ahead feature.

For the SolarTAC data set, we consider 3 weeks data for training. The data uses 15-minute intervals and includes only October 2016. Hence a little over 1 week is

Table 6.6
Pinball loss average for using 4 approaches in Elia, Golden CO, and Aurora CO

Approach	Elia Data	SRRL Data	SolarTAC Data
XGBoost Quantile	87.7268	0.1007	39.4606
Conformal	86.3896	0.1015	22.0536
Gaussian Baseline	86.8460	0.1019	22.9608
Naive Hourly	120.6003	0.1022	50.6377

used in testing. The features include five lag values including the previous 15, 30, and 45 minutes as well as 1 day and 2 days in advance. In addition, it also includes the data from SRRL which is 30 minutes away by including the unscaled global value and its air temperature. We also use the air temperature at SolarTAC, hour, and minute values as features. Then we use it to predict the global solar irradiance (the unscaled version). All the XGBoost parameters are the same as what was used in Elia.

Refer to Table 6.6. It is important to not compare across the locations as they are very different types of data, at different granularity, some scaled and some unscaled, and with different features. In all methods, as expected, naive hourly is the worst. While that is expected, the main reason it is used is to see how much better the others are. In case of SRRL there is not a huge difference, although it is the worst. That is perhaps because we only consider 1 week to train for a 3 month data set. So all the data and perhaps all fairly random. Of course the values themselves are small, that is because of using scaled irradiance.

Since we used the same XGBoost parameters in all the problems, it would certainly be better if one were to redesign the XGBoost. Nonetheless, it appears like for Elia and SolarTAC data, the conformal approach is the best. Whereas for the SRRL data, XGBoost quantile performed the best. It is important to note that the Gaussian forecast lags the conformal predictions in all three data sets, but not by much. Whereas the XGBoost Quantile is a little more away from conformal. In summary, all methods performed well and it is not clear if one of them is necessarily better. But it would be useful to reconsider the XGBoost parameters and the data before finalizing the preferred one.

CONCLUDING REMARKS

As we draw to the end of this book, there are a few things to remember. First, although there are methods that have been compared here and in previous chapters, it is not that one is always better than the other. The point is to show the reader how one would go about making forecasts. Then it is up to the reader to try various methods and decide for themselves which might work best. Second, not all methods have been described here. In fact, there are a vast number of neural network based methods that have not been discussed. That would be useful to consider before making a final choice. Third, it is not possible to get the perfect forecast and one has to live with these probabilistic forecasts. This is especially true for solar power.

The topics of forecasting (predictive analytics) and deciding how much to buy from the grid (prescriptive analytics) have been well studied. There have been numerous articles written about them. But the topic of probabilistic forecasts have not received nearly that level of attention. What it really brings is that the observed ground truth itself is a random event. Hence we really should not be taking that as given but must provide some level of uncertainty to it. This is especially true with solar power and related forecasts as they depend so much on weather. There is only a little that can be captured as data that can affect the choice. So, there is a lot of uncertainty and that may be best left as is.

That said, the author wishes to conclude this book by providing a hope in the near future. Everyone in this world would have access to electricity, and a vast majority through solar and wind. The electricity is clean and safe. There are batteries that are used at times when solar or wind is not available. Analytics engines do predictions through various forecasting techniques so that the systems can be used as efficiently as possible. While this is a dream, we are not far from it becoming a reality. The hope is that it would come soon. Cheers!

References

1. Mohammadali Alipour, Jamshid Aghaei, Mohammadali Norouzi, Taher Niknam, Sattar Hashemi, and Matti Lehtonen. A novel electrical net-load forecasting model based on deep neural networks and wavelet transform integration. *Energy*, 205:118106, 2020.
2. Afshin Andreas and Tom Stoffel. Solar Radiation Research Laboratory (SRRL): Baseline Measurement System (BMS); Golden, Colorado (Data). *NREL Report No. DA-5500-56488*, 2011.
3. Afshin Andreas and Steve Wilcox. Solar Technology Acceleration Center (SolarTAC) Aurora, Colorado (Data). *NREL Report No. DA-5500-56491*, 2011.
4. Afshin Andreas and Steve Wilcox. Solar Resource & Meteorological Assessment Project (SOLRMAP): Rotating Shadowband Radiometer (RSR); Los Angeles, California (Data). *NREL Report No. DA-5500-56502*, 2012.
5. Khadija Barhmi, Chris Heynen, Sara Golroodbari, and Wilfried van Sark. A review of solar forecasting techniques and the role of artificial intelligence. In *Solar*, pages 99–135. MDPI, 2024.
6. Belgium Elia Group. Website: `https://opendata.elia.be/pages/home/`.
7. Frede Blaabjerg, Yongheng Yang, Katherine A. Kim, and Jose Rodriguez. Power electronics technology for large-scale renewable energy generation. *Proceedings of the IEEE*, 111(4):335–355, 2023.
8. Subhonmesh Bose and Eilyan Bitar. Variability and the locational marginal value of energy storage. In *53rd IEEE Conference on Decision and Control*, pages 3259–3265. IEEE, 2014.
9. Tianqi Chen. Website: `https://www.youtube.com/watch?v=Vly8xGnNiWs`.
10. Tianqi Chen and Carlos Guestrin. Xgboost: A scalable tree boosting system. In *Proceedings of the 22nd ACM SIGKDD international conference on knowledge discovery and data mining*, pages 785–794, 2016.
11. M Hisham Choueiki, Clark A Mount-Campbell, and Stanley C Ahalt. Implementing a weighted least squares procedure in training a neural network to solve the short-term load forecasting problem. *IEEE Transactions on Power Systems*, 12(4):1689–1694, 2002.
12. Yu Ding. *Data science for wind energy*. Chapman and Hall/CRC, 2019.
13. John A. Duffie and William A. Beckman. *Solar engineering of thermal processes*. Wiley New York, 1980.
14. Javier Garcia-Gonzalez, Rocio R. de la Muela, Luz M. Santos, and Alicia M. Gonzalez. Stochastic joint optimization of wind generation and pumped-storage units in an electricity market. *IEEE Tran. Power Systems*, 23(2):460–468, 2008.
15. Trevor Hastie. Website: `https://www.youtube.com/watch?v=wPqtzj5VZus`.
16. Luis Hernandez, Carlos Baladron, Javier M Aguiar, Belen Carro, Antonio J. Sanchez-Esguevillas, Jaime Lloret, and Joaquim Massana. A survey on electric power demand forecasting: future trends in smart grids, microgrids and smart buildings. *IEEE Communications Surveys & Tutorials*, 16(3):1460–1495, 2014.
17. Henrique Steinherz Hippert, Carlos Eduardo Pedreira, and Reinaldo Castro Souza. Neural networks for short-term load forecasting: A review and evaluation. *IEEE Transactions on Power Systems*, 16(1):44–55, 2002.

18. Hui Hou, Chao Liu, Qing Wang, Xixiu Wu, Jinrui Tang, Ying Shi, and Changjun Xie. Review of load forecasting based on artificial intelligence methodologies, models, and challenges. *Electric Power Systems Research*, 210:108067, 2022.
19. Longbo Huang, Jean Walrand, and Kannan Ramchandran. Optimal demand response with energy storage management. In *IEEE International Conference on Smart Grid Communications (SmartGridComm)*, pages 61–66, 2012.
20. Rob J. Hyndman and George Athanasopoulos. *Forecasting: Principles and Practice, 3rd Edition.* OTexts Melbourne, 2021. `https://otexts.com/fpp3/`
21. Rob J. Hyndman, Anne B. Koehler, Ralph D. Snyder, and Simone Grose. A state space framework for automatic forecasting using exponential smoothing methods. *International Journal of Forecasting*, 18(3):439–454, 2002.
22. Rich H. Inman, Hugo T.C. Pedro, and Carlos F.M. Coimbra. Solar forecasting methods for renewable energy integration. *Progress in Energy and Combustion Science*, 39(6):535–576, 2013.
23. David Jager and Afshin Andreas. NREL National Wind Technology Center (NWTC). *M2 Tower, Boulder Colorado (Data); NREL Report No. DA-5500-56489*, 1996.
24. Jan Kleissl. *Solar energy forecasting and resource assessment.* Academic Press, 2013.
25. Rok Koritnik. *At the turning of years: A book on Solar forecasting.* Self Published, 2022.
26. Iordanis Koutsopoulos, Vasiliki Hatzi, and Leandros Tassiulas. Optimal energy storage control policies for the smart power grid. *Proc. of IEEE International Conference on Smart Grid Communications (SmartGridComm)*, 2011.
27. Soongeol Kwon, Lewis Ntaimo, and Natarajan Gautam. Optimal day-ahead power procurement with renewable energy and demand response. *IEEE Transactions on Power Systems*, 32(5):3924–3933, 2016.
28. Soongeol Kwon, Lewis Ntaimo, and Natarajan Gautam. Demand response in data centers: Integration of server provisioning and power procurement. *IEEE Transactions on Smart Grid*, 10(5):4928–4938, 2018.
29. Soongeol Kwon, Yunjian Xu, and Natarajan Gautam. Meeting inelastic demand in systems with storage and renewable sources. *IEEE Transactions on Smart Grid*, 8(4):1619–1629, 2015.
30. Subhash Lakshminarayana, Tony Q.S. Quek, and H. Vincent Poor. Cooperation and storage tradeoffs in power grids with renewable energy resources. *IEEE Journal on Selected Areas in Communications*, 32(7):1386–1397, 2014.
31. Tiantian Li, Bo Wang, Lianming Zhang, Xin Zhao, et al. Short-term load forecasting using optimized lstm networks based on emd. In *2018 10th International Conference on Communications, Circuits and Systems (ICCCAS)*, pages 84–88. IEEE, 2018.
32. Jean H. Meeus. *Astronomical algorithms.* Willmann-Bell, Incorporated, 1991.
33. NASA Irradiance. Website: `https://www.nasa.gov/mission_pages/sdo/science/Solar%20Irradiance.html`.
34. NREL Solar Data. Website: `http://www.nrel.gov/midc/`.
35. Sophie Pelland, Jan Remund, Jan Kleissl, Takashi Oozeki, and Karel De Brabandere. Photovoltaic and solar forecasting: state of the art. *IEA PVPS Task*, 14(355):1–36, 2013.
36. Hongqiao Peng, Yuanyuan Lou, Feng Li, Hui Sun, and Ruikuan Liu. Long-term regional hourly net load profile generation with behind-the-meter PV. In *2024 IEEE PES 16th Asia-Pacific Power and Energy Engineering Conference (APPEEC)*, pages 1–5. IEEE, 2024.

37. Nicholas C. Petruzzi and Maqbool Dada. Pricing and the newsvendor problem: A review with extensions. *Operations Research*, 47(2):183–194, 1999.
38. Ravid Shwartz-Ziv and Amitai Armon. Tabular data: Deep learning is not all you need. *Information Fusion*, 81:84–90, 2022.
39. George Sideratos, Andreas Ikonomopoulos, and Nikos D Hatziargyriou. A novel fuzzy-based ensemble model for load forecasting using hybrid deep neural networks. *Electric Power Systems Research*, 178:106025, 2020.
40. Pardeep Singla, Manoj Duhan, and Sumit Saroha. A comprehensive review and analysis of solar forecasting techniques. *Frontiers in Energy*, 16(2):187–223, 2022.
41. Arno Smets, Klaus Jäger, Olindo Isabella, Rene Van Swaaij, and Miro Zeman. *Solar Energy: The physics and engineering of photovoltaic conversion, technologies and systems.* Bloomsbury Publishing, 2016.
42. Anne Stratman, Tianqi Hong, Ming Yi, and Dongbo Zhao. Net load forecasting with disaggregated behind-the-meter PV generation. *IEEE Transactions on Industry Applications*, 59(5):5341–5351, 2023.
43. Confronting the Duck Curve: How to Address Over-Generation of Solar Energy. Website: `https://www.energy.gov/cmei/solar/articles/confronting-duck-curve-how-address-over-generation-solar-energy`.
44. Philip E. Tetlock and Dan Gardner. *Superforecasting: The art and science of prediction.* Random House, 2016.
45. Aidan Tuohy, John Zack, Sue Ellen Haupt, Justin Sharp, Mark Ahlstrom, Skip Dise, Eric Grimit, Corinna Mohrlen, Matthias Lange, Mayte Garcia Casado, et al. Solar forecasting: methods, challenges, and performance. *IEEE Power and Energy Magazine*, 13(6):50–59, 2015.
46. John Twidell. *Renewable energy resources.* Routledge, 2021.
47. Venkata R. Vakacharla, K. Gnana, P. Xuewei, B.L. Narasimaharaju, Mangu Bhukya, Atanu Banerjee, Renu Sharma, and Akshay K. Rathore. State-of-the-art power electronics systems for solar-to-grid integration. *Solar Energy*, 210:128–148, 2020.
48. Cyril Voyant, Gilles Notton, Soteris Kalogirou, Marie-Laure Nivet, Christophe Paoli, Fabrice Motte, and Alexis Fouilloy. Machine learning methods for solar radiation forecasting: A review. *Renewable Energy*, 105:569–582, 2017.
49. Can Wan, Jian Zhao, Yonghua Song, Zhao Xu, Jin Lin, and Zechun Hu. Photovoltaic and solar power forecasting for smart grid energy management. *CSEE Journal of Power and Energy Systems*, 1(4):38–46, 2015.
50. Yi Wang, Ning Zhang, Qixin Chen, Daniel S Kirschen, Pan Li, and Qing Xia. Data-driven probabilistic net load forecasting with high penetration of behind-the-meter PV. *IEEE Transactions on Power Systems*, 33(3):3255–3264, 2017.
51. What the Duck Curve Tells us about Managing a Greed Grid. Website: `https://www.caiso.com/Documents/FlexibleResourcesHelpRenewables_FastFacts.pdf`.
52. Adam Wierman, Zhenhua Liu, Iris Liu, and Hamed Mohsenian-Rad. Opportunities and challenges for data center demand response. In *Green Computing Conference (IGCC), 2014 International*, pages 1–10. IEEE, 2014.
53. Le Xie, Yingzhong Gu, Ali Eskandari, and Mehrdad Ehsani. Fast MPC-based coordination of wind power and battery energy storage systems. *Journal of Energy Engineering*, 138(2):43–53, 2012.
54. Gokhan Mert Yagli, Dazhi Yang, and Dipti Srinivasan. Automatic hourly solar forecasting using machine learning models. *Renewable and Sustainable Energy Reviews*, 105:487–498, 2019.

55. Dazhi Yang, Wenting Wang, Christian A. Gueymard, Tao Hong, Jan Kleissl, Jing Huang, Marc J. Perez, Richard Perez, Jamie M. Bright, Xiang'ao Xia, et al. A review of solar forecasting, its dependence on atmospheric sciences and implications for grid integration: Towards carbon neutrality. *Renewable and Sustainable Energy Reviews*, 161:112348, 2022.
56. Zhaoyun Zeng, Paul Stackhouse, Ji-Hyun Jeannie Kim, and Ralph T. Muehleisen. Development of typical solar years and typical wind years for efficient assessment of renewable energy systems across the us. *Applied Energy*, 377:124698, 2025.
57. Qingyong Zhang, Jiahua Chen, Gang Xiao, Shangyang He, and Kunxiang Deng. Transformgraph: A novel short-term electricity net load forecasting model. *Energy Reports*, 9:2705–2717, 2023.
58. Tingze Zhang, Xinan Zhang, San Shing Choi, Tat Kei Chau, Yau Chow, Tyrone Fernando, and Herbert Ho-Ching Iu. Highly accurate peak and valley prediction short-term net load forecasting approach based on decomposition for power systems with high PV penetration. *Applied Energy*, 333:120641, 2023.
59. Yating Zhou and Meng Wang. Unifying load disaggregation and prediction for buildings with behind-the-meter solar. *IEEE Transactions on Power Systems*, 2024.

Index

For Product Safety Concerns and Information please contact our EU
representative GPSR@taylorandfrancis.com
Taylor & Francis Verlag GmbH, Kaufingerstraße 24, 80331 München, Germany

www.ingramcontent.com/pod-product-compliance
Ingram Content Group UK Ltd.
Pitfield, Milton Keynes, MK11 3LW, UK
UKHW020203200926
475702UK00004B/598
* 9 7 8 1 0 3 2 5 1 5 3 2 8 *